Charles Bryant

Flora diaetetica

History of esculent plants, both domestic and foreign

Charles Bryant

Flora diaetetica
History of esculent plants, both domestic and foreign

ISBN/EAN: 9783337271763

Printed in Europe, USA, Canada, Australia, Japan

Cover: Foto ©berggeist007 / pixelio.de

More available books at **www.hansebooks.com**

FLORA DIÆTETICA:
OR,
HISTORY
OF
ESCULENT PLANTS,
Both DOMESTIC and FOREIGN.

IN WHICH

They are accurately defcribed, and reduced to their LINNÆAN Generic and Specific Names.

WITH

Their ENGLISH NAMES annexed, and ranged under Eleven GENERAL HEADS,

VIZ.

ESCULENT
1 ROOTS,
2 SHOOTS, STALKS, &c.
3 LEAVES,
4 FLOWERS,
5 BERRIES,
6 STONE-FRUIT,
7 APPLES,
8 LEGUMENS,
9 GRAIN,
10 NUTS,
11 FUNGUSES.

AND

A particular Account of the Manner of ufing them; their native Places of Growth; their feveral Varieties, and Phyfical Properties: Together with whatever is otherwife curious, or very remarkable in each Species.

THE WHOLE

So methodized, as to form a fhort INTRODUCTION to the SCIENCE OF BOTANY.

By CHARLES BRYANT, of *Norwich.*

LONDON:
Printed for B. WHITE, at Horace's Head, in Fleet-ftreet.
M.DCC.LXXXIII.

T O

JAMES CROWE, Efq;

O F

TUCK's WOOD, NEAR NORWICH,

THIS HISTORY

O F

ESCULENT PLANTS

Is, with all due Submiffion, Infcribed,

B Y

His moft Obedient

Humble Servant,

C. BRYANT.

PREFACE.

WHETHER we view Mankind in a natural or civilized ſtate, we ſhall find that the principal part of his daily food, and alſo moſt of the articles neceſſary to his comfortable enjoyment of Life, are drawn from the vegetable kingdom ; every endeavour therefore to point out with preciſion and accuracy the Species of Plants, immediately adapted to the uſe of man, muſt carry with it its own recommendation ; for, by furniſhing him with the means of diſtinguiſhing the different Species of plants clearly, he is thereby enabled to chooſe ſuch as are moſt wholeſome, and beſt ſuited to his palate and conſtitution, and of rejecting ſuch as are diſagreeable and hurtful. Now this can never be anſwered by any method ſo well, as

by

by that of calling plants after their generic and trivial Names, for thefe once acquired, any particular Species may be as certainly difcourfed upon, as any fingle letter in the Alphabet. By thefe, Botany is reduced to a permanent and univerfal language, which may be adopted by all people and nations ; but without thefe, the moft laboured defcriptions often prove ineffectual, and the meaning liable to be miftaken. The truth of this is evident from the writings of many travellers, who have endeavoured to defcribe the plants peculiar to the feveral countries they have paffed through ; but though they have taken much pains to be underftood, yet it is frequently out of the power of the moft expert Botanift, to be certain of many plants they mention, for want of their defcriptions being delivered according to the language of Botany ; or, if the plants were fuch

as

as are in Linnæus, their not speaking
of them by their generic and trivial
names. These names would be all
the descriptions neceffary to a scien-
tific Botanist, and this method
would fave fuch travellers a great
deal of time, but for want of proceeding in this way, their labours
become almoft ufelefs, and the œconomy of human life is often robbed
of many advantages. Hence, a-
mongft other inftances, Botany becomes a fcience of the firft confequence, and claims the moft liberal
encouragement, as when it is properly underftood and applied, it
may be productive of the greateft benefits to mankind. All Gentlemen
then that travel with the public good
in view, fhould previoufly acquire
fuch a ftock of the Linnæan fyftem
as will enable them to reduce plants
to their Genera and Species. Nor
is a competent knowledge of this
Science lefs neceffary to the ftationed
A 4   Gentleman ;

Gentleman; for furely it ill fuits with the character of a perfon of a polite education, to adopt the vulgarifms of the unlearned. And yet for the moft part this is the cafe, there being nothing more frequent than for people in a high ftation of life, to converfe about their fruits and fallads, under the barbarous names they may have heard them called by, and which are often local. Gardeners and Nurferymen too ought to be well acquainted with the Linnæan names of the plants they cultivate and deal in, the want of which knowledge many times renders their language unintelligible even among themfelves, efpecially if they have been brought up in different places. The utility of the following Manual then muft immediately appear, as by it any one may furnifh himfelf with the Linnæan names of moft of the efculent plants in ufe throughout the known

parts

parts of the globe, and that with
very little trouble; for it being
portable in the pocket, and fufficient in itfelf for the purpofe, Gentlemen and Ladies not at all acquainted with Botany, may amufe
themfelves in their gardens, and ex-
amine the greateft part of their vegetables fcientifically, without the
fatigue of regularly ftudying the
Science, as all. fuch terms as were
unavoidable in true defcription, are
explained at the beginning of. the
work. Under this view, likewife,
it muft become directly ufeful to
thofe who travel, as they will be
hereby enabled to fatisfy themfelves
in regard to the edible plants they
may meet with abroad, and in their
writings be capable of giving the
country names in conjunction with
the true botanical ones, a thing of
no fmall confequence in Hiftory.

Some time paft, Mr. Hugh Rofe,
Apothecary of Norwich, for his
own

own information, fet himfelf about
collecting the Linnæan names of the
Efculent Plants ; his lift coming into
my hands, I made as many additions
to it as I could, have defcribed all
the plants, except fuch as are generally known ; and have digefted
and divided the whole into eleven
general Heads, with Subdivifions of
them, that the defcriptions of the
plants might immediately follow
every fmall parcel of names. Thefe
defcriptions I have delivered in as
plain and fimple language as poffible,
being fenfible, that a work chiefly
intended to bring into general ufe
the fcientific names of a particular
fet of plants only, could not be expreffed in too familiar a ftyle. I
have likewife aimed at brevity, as
well as plainnefs, not unfrequently
making one plant fubfervient to the
defcription of another, by only con-
trafting their difference. As to fuch
Exotics as could not come under my
infpection,

infpection, I have deliberately con-
fulted the beft botanic writers upon
them, and by comparing their feveral defcriptions, have formed fuch
as I hope will be found to give the
moft accurate ideas of the plants
defcribed. Knowing alfo, that many
readers are very folicitous about the
virtues of plants, I have added the
moft general phyfical properties of
the greateft part of thefe, as far as
could, with propriety, be deduced
from their material compofitions,
and perceivable effects upon the organs of fenfation.

Having pointed out the principal
defign, it remains to mention but
one circumftance more refpecting
the fucceeding pages ; which is,
that feveral plants inferted there
were never yet generally introduced
into the kitchen, but all of them
have been privately tried, and found
to be equal, nay even to furpafs many,
whofe ufes have been long eftablifhed.

ed.   This muſt prove of public advantage, in particular ſeaſons, as out
of ſuch a number, if ſome ſhould
fail, others will be in perfection;
and ſurely no one will object to
increaſing the Eſculent Plants, from
an opinion of its tending to promote
luxury, eſpecially if he reflect that
human health and vigour can never
be ſupported ſo well, as by a frequent uſe of vegetable diet, and that
by having a great variety to chooſe
from, both the palates and pockets
of different people will be the more
agreeably accommodated.

TERMS

TERMS EXPLAINED.

AS it was impossible to deliver a work of this kind, with any tolerable propriety, without making use of some of the terms peculiar to the science, it will be necessary for such readers as may be entire strangers to Botany, to get a perfect idea of the few general ones following, before they consult the descriptions of the plants, otherwise they will not be able clearly to comprehend them, as these words are constantly occurring.

1 Annual.	15 Calyx.
2 Biennial.	16 Catkin.
3 Perennial.	17 Petal.
4 Sessile.	18 Glume.
5 Serrated.	19 Arista, or Awn.
6 Crenated.	20 Floret.
7 Pinnated, or winged.	21. Germen, or Seed-bud.
8 Peduncle.	22 Pericarpium.
9 Spike.	23 Capsule.
10 Spicula.	24 Stamina.
11 Panicle.	25 Styles.
12 Spadix.	26 Stigma.
13 Racemus.	27 Summit.
14 Umbel.	

1 A plant is said to be *Annual* when it dies, root and branch, in the course of the year in which it vegetated; as Common Barley.

2 A *Biennial* plant is that which totally perishes the second year after it vegetated; as Garden Clary.

3 A *Pere-*

3 A *Perennial* plant is fuch whofe root continues alive in the ground for many years; as feveral forts of Mint.

4 A leaf or flower is faid to be *Seffile* when it has no foot-ftalk; as the leaf and flower of the Garden Purflane.

5 A *ferrated* leaf is fuch as hath its margin cut into teeth like thofe upon the edge of a faw; as in the Rofe leaf.

6 A leaf is faid to be *crenated* when its margin is cut into femicircular teeth; as the leaf of Ground Ivy.

7 A *pinnated* or *winged* leaf hath feveral leffer leaves placed on each fide a common foot-ftalk; as the leaf of the Afh.

8 A *Peduncle* is the ftalk that fupports a flower, and is fo called to diftinguifh it from the ftalk that fuftains a leaf.

9 A *Spike* is formed by many feffile flowers ftanding on both, or on all fides a common peduncle*; as a fpike of Lavender.

10 A *Spicula* is a partial fpike; in Wheat, the main fpike is compofed of a number of fpiculæ.

11 A *Panicle* is formed by the flowers being varioufly branched from the extremity of a common ftem, upon feparate peduncles; as in the Oat.

12 A *Spadix* is a flower-ftem that is protruded out of a fheath; as that of the Common Arum.

13 A *Racemus* is a long bunch of flowers, each

---

* This fort of peduncle Linnæus more properly calls a *Receptacle.*

of

of which is fupported on a diftinct peduncle, which fprings from the fide of a common peduncle; as in the Currants.

14 An *Umbel* is a bunch of flowers, in which many common peduncles, rifing to an equal height, proceed from a point at the extremity of a ftem, and fupport the flowers in fmall clufters; as in Common Parfley.

15 A *Calyx* is the leaf or leaves that enclofe and protect the other parts of a flower before they expand.

16 A *Catkin* is a fort of compound calyx, confifting of a great many fcales, ranged along a common receptacle, and has obtained its name from its refembling a cats-tail; as in the Willow.

17 The *Petal* or *Petals* of a flower, are the leaf or leaves placed within the calyx, and are of various fhapes and colours, according to the nature of the plant.

18 *Glume* * is a term which ought to be applied only to grafs-leaved plants, and fhould be confined to point out the chaffy leaf that immediately furrounds the feed.

19 An *Arifta* or *Awn*, is a fort of beard that fprings from fome part of a hufk or feed of the grafs-leaved plants; as the beard of Barley.

20 A *Floret* is a partial flower; a compleat

---

* *Gluma*, a *hufk*, has not hitherto had any definite meaning in Botany, which has caufed fome confufion even in the works of Linnæus; for in defcribing the grafs-leaved plants, he fometimes ufes it for the calyx, and fometimes for the petals, or chaff that furround the feed, whereby it is not always poffible to underftand his meaning. The propriety of confining it to the petals therefore muft immediately appear.

flower

flower of the Dandelion is compofed of a number of florets.

21 The *Germen* is the rudiment of the fruit, or feed-veffel.

22 A *Pericarpium* is a feed-veffel arrived at maturity.

23 A *Capfule* is a dry hollow feed-veffel, that cleaves or fplits in fome certain manner; as a Poppy-head.

24 The *Stamina* are the little threads ftanding within the petals, and are called the male organs of generation.

25 The *Styles* are fmall pillars, moftly placed in the centre of the ftamina, and are the female organs of generation.

26 The *Stigma* is the top of the ftyle, and is varioufly formed.

27 The *Summits* are the tops of the ftamina.

HISTORY

OF

ESCULENT PLANTS.

CHAPTER I.

ESCULENT ROOTS.

SECTION I.

Roots now or formerly made use of as Bread.

1 ARUM colocasia. *Egyptian Arum or Colocasia.*
2 Arum esculentum. *Eatable Arum.*
3 Arum peregrinum. *Edders.*
4 Calla palustris. *Water Dragons.*
5 Convolvulus batatas. *Spanish Potatoes.*
6 Dioscorea sativa. ⎫
7 Dioscorea alata. ⎬ *Indian Yams.*
8 Dioscorea bulbifera. ⎭
9 Jatropha maniot. *Cassava or Indian Bread.*
10 Nymphæa lotus. *Egyptian Lotus.*
11 Sagittaria sagittifolia. *Common Arrowhead.*
12 Solanum tuberosum. *Common Potatoes.*
13 Yucca gloriosa. *Adam's Needle.*

B   14 Polygonum

14 Polygonum divaricatum. *Eastern Buckwheat*.

1 ARUM colocasia. *Lin. Sp. pl.* 1368.
Arum maximum Egyptium, quod vulgo
Colocasia. *Bauh. Pin.* 195.

Great has been the controversy amongst
ancient Botanists concerning this-plant; some
insisting that it was the *Faba Egyptia* of
Dioscorides and Theophrastus, and others
denying it, contending with good reason that
it was the seed of the *Faba Egyptia* that
was eaten, and not the root. This plant no
doubt is the true *Colocasia* of the ancients,
and the same which is mentioned by Virgil
in his Eclogues *. It grows in Crete, Cy-
prus, Syria, and Egypt, propagating itself
chiefly by its roots; for it flowers so late,
that it can perfect its seeds only in particular seasons. This last circumstance induced
many travellers to believe it was not natural
to these parts, but had been introduced there,
and was the means of leading them into
mistakes about the plant, the general habit
of which somewhat agreeing with that of
the *Faba Egyptia*, and some asserting that
the root of the latter was eaten, they im-
plicitly pronounced the former to be the
Faba Egyptia, the root of which had been
affirmed by some to be the true *Colocasia*.

Dr. Hasselquist met with the *Arum colocasia* both in the fields and gardens of Egypt.

* Eclog. iv. v. 20.

It

It hath a large tuberous root covered with a brownifh fkin, but when cut is white within, and of a fharpifh acrid tafte.

The leaves come immediately from the root on long, thick footftalks; they are large, and fomewhat of the fhape of thofe of the Butter bur, of a dark fhining green colour, and have their footftalk inferted near their centre.

Among the leaves rifes the flower-ftalk, which is round, of a pale green, and terminated by a large fheath including a peftle, or clapper, like that of our Wake-Robin, but longer, thinner, and fet round at the bottom with red berries.

The roots of moft of the fpecies of this genus are intolerably acrimonious, but this is of a milder nature, and much efteemed by the inhabitants of the Eaft for its nutritious quality. What pungency it has is taken out by foaking it in water for fome hours, after which, it is dried and is then fit for table. Sometimes, however, they are boiled or roafted, and eaten as potatoes. A root or two of *Colocafia* with a glafs of good wine is a pleafant regale.

2 ARUM efculentum. *Eatable Arum. Lin. Sp. pl.* 1369.

Arum minus, nymphææfolio, efculentum. *Sloan. Jam.* 62.

This is a native of America. It is a much fmaller plant than the former, and

B 2　　　　has

has leaves refembling our Water Lily. The inhabitants of the fugar iflands cultivate it in plenty, as food for their flaves. It has a mild root, and not only this is eaten, but the leaves alfo, which are a favourite fallad among the Indians, and on that account they are called Indian Kale. This circum-ftance probably induced Linnæus to give it the trivial name of *efculentum*, the better to diftinguifh it from thofe *Arums*, whofe roots only are eaten.

3 ARUM peregrinum. *Edders. Lin. Sp. pl.* 1369.

This is likewife a native of America, and is cultivated for the roots in the fame man-ner as that juft mentioned. It differs from the *efculentum*, in having leaves between the form of an heart and that of a fpear. The roots of both the fpecies are eaten the fame as are potatoes with us, and the Edders are very pleafant.

There are fome others of this genus, whofe roots are efculent, as thofe of the *fagittifolium*, but they are not fo generally cultivated.

4 CALLA paluftris. *Water Dragons. Lin. Sp. Pl.* 1373.

Dracunculus aquatilis. *Dod. plant.* 330.

The roots of this are faid to be eaten, but in what manner I cannot learn. It is a native

tive of the northern parts of Europe, and
grows in the marfhes. The root is thick,
flefhy, and jointed. It creeps in the mud,
and fends up in clufters many fiftulous ftalks,
fupporting heart-fhaped, deep green leaves.
The flower-ftems rife in the midft of the
tufts of leaves, to about eight inches high;
they are round, thick, of a pale green, and
are furrounded by the bafes of the leaves.
Each ftem terminates with a light green,
plain fpatha, which is fnipped at its bafe,
and includes a club-fhaped fpadix, furrounded with hermaphrodite, whitifh, chivy
flowers, having neither calyx, nor petals,
but are fucceeded by red globular berries,
ftanding round the fpadix, as they do in
common Arum. The fpatha is permanent,
and remains with the fruit.

5 Convolvulus batatas. *Spanifh Pota-
toes. Lin. Sp. pl. 220.*
· Convolvulus indicus vulgo Patates dictum.
Raii Hift. 728.

The *Batatas* is a native of both Indies,
but has been a long time cultivated in Spain
and Portugal, whence the roots are annually
imported.

It puts forth many long, trailing ftalks,
which are very rough, and as they run on
the ground they ftrike fibres, and produce
large, irregular, tuberous roots. The ftalks
are furnifhed with almoft fpear-fhaped

leaves,

leaves, of a dark green colour, with five prominent veins running through each. The flowers are produced at the bofoms of the leaves, on long peduncles; they are bell-fhaped, fpread open at the top, and contain five ftamina and one ftyle each, crowned with a forked ftigma.

The root is firm, of a pale brown on the outfide, white within, very fweet, and is the only one at prefent known, in all this copious genus, to be efculent, thofe of the reft of the fpecies being either very pungent or violently cathartick. It is a plant that well repays the time and labour of the cultivators, for one bufhel of the roots generally yields fifty; but we cannot reap this benefit, as our climate is not warm enough to produce the plant to perfection.

Thefe are certainly the fame fpecies of roots as thofe which Columbus's failors were treated with by the inhabitants of Cuba, and which they faid were very fweet, and when boiled tafted like chefnuts.

6 DIOSCOREA fativa. *Yams. Lin. Sp. pl.* 1463.

Volubilis nigra, folio cordato nervofo. *Sloane Jam.* 46. *Hift.* I. *p.* 140.

This is a native of both the Indies, and is cultivated in all the fugar iflands in the Weft, where the roots are the principal food of the Negroes.

It

It sends forth many weak, smooth, slender stalks, which fix themselves to any support near them, in the manner of our Briony, some of them running to the length of twenty feet; they are blackish, are furnished with heart-shaped leaves, ending in acute points, and each has five longitudinal veins, which take their rise at the base, and diverge towards the sides, but meet again at the apex. The flowers come out in a racemus at the footstalks of the leaves; they have no petals, but consist of a small calyx cut into six parts, and are male and female in distinct plants. The male flower has six hairy stamina, and the female a three cornered germen, crowned with three styles, and becomes a capsule of three cells, each containing two membranous seeds.

7 DIOSCOREA alata. *Yams. Lin. Sp. pl.* 1462.

Volubilis rubra, caule membranulis extantibus alato, folio cordato nervoso. *Sloan. Jam.* 46. *Hist.* I. *p.* 140.

This too grows spontaneously in both the Indies, and is cultivated in manner of the former. It differs from the *sativa* in being a smaller plant, in the stalks being red, triangular, and winged, and sometimes putting out bulbs at their joints, as they trail on the ground.

8 DIOS-

8 Dioscorea bulbifera. *Yams. Lin. Sp. pl.* 1463.

Rhizophora Zeylanica, fcammonii folio fingulari, radice rotunda. *Herm. Par.* 217. *t.* 217.

This differs from both the former in the roots being rounder. Its leaves refemble thofe of Scammony in their fhape, but they are warted.

The roots of all thefe three fpecies are promifcuoufly eaten, by the name of *Yams;* they differ greatly in colour, fize, and fhape; fome being blueifh, fome brownifh; and as to fhape, fome are round, others irregularly oblong. With refpect to fize, they weigh from a pound to ten and upwards. They are of a very nutritious nature, eafy to digeft, and when dreffed, are preferred to the beft wheaten bread. The tafte is fomewhat like the potatoe, but more lufcious. For negroe food they are generally boiled, and then beaten into a mafh. The white people grind them to flour, and make bread and puddings of them. In order to have the benefit of them the year through, upon digging them up, they are expofed in the fun to dry, in the manner of our onions, and when fufficiently weathered, they are preferved in dry fand, garrets, or cafks, and if kept from moifture, will continue feveral feafons, and lofe nothing of their primitive goodnefs.

9 Jatropha

9 JATROPHA maniot. *Caſſava. Lin.
Sp. pl.* 1429.

Arbor ſucco venenato, radice eſculenta.
Baub. Pin. 512.

The *Caſſava* is a native of the warmer
parts of America. It is a ſhrubby plant,
ſending up ſeveral ſtalks ſeven or eight feet
high, which are covered with a thin bark,
of different hues, according to the age of
the ſtems, it being grey, red, or blue. The
ſtalks and branches are furniſhed with
ſmooth, hand-ſhaped leaves, conſiſting of
five or ſeven lance-ſhaped lobes each. The
flowers come out in bunches at the tops of
the ſtalks, ſome being male and others female. The male has no calyx, but is compoſed of a bell-ſhaped petal, containing ten
ſtamina, forming a column. The female
alſo has no calyx, and conſiſts of five whitiſh
petals, ſurrounding three bifid ſtyles, and
is ſucceeded by a capſule of three cells, containing one ſeed each. The principal root
is about half a yard long, and two or three
inches thick; almoſt cylindrical, red or
greyiſh on the outſide, white within, of a
farinaceous ſubſtance, mixed with a milky
juice, and every part of it is a fatal poiſon
when raw; but notwithſtanding this, theſe
roots furniſh a very great part of the daily
food of the inhabitants of all denominations
in the Weſt Indies.

When theſe roots are full grown and fit
for uſe, it requires no great labour to get
them

them up, for they do not penetrate far into the ground, and therefore the method ufed by the negroes, is, to pluck up the whole tree, roots and all, and if any of the offsets chance to feparate, which is fometimes the cafe, they draw thefe up with a hoe. In order to prepare them for food, they pare off the outer bark with a coarfe knife; then the roots are rubbed on large copper graters to reduce them to meal, which much re-fembles the fawings of fome white grained wood. When a fufficient quantity of meal is obtained, it is put into a prefs, and the watery part fqueezed from it, and carefully fet by in veffels kept at hand for the purpofe. The fubftantial part is then taken from the prefs, and if immediately wanted for bread, it is made into cakes, and baked upon iron plates over a flow fire, till they become brown; after which they will keep fweet for feveral months. The plates are about two feet broad, and half an inch thick, and are placed either upon ftones, or an iron trivet. A fire is made underneath, and when the iron is properly heated, which they try by touching it with their fingers, they lay the meal on equally over the whole plate, till they have covered it about two inches thick. As it roafts, the perfon that attends it, gently paffes a fmooth piece of wood over the furface, which caufes the mafs to incorporate and fubfide, till it be-comes not above the eighth of an inch thick.

When

When baked enough, it is taken off, and laid a few hours in the fun, that if any moifture yet remains, it may be diffipated, and thereby prevent the cake from contracting a mould. This bread is eafy to digeft, very nourifhing, though but coarfe in the mouth. A piece of it dipped in water or other liquor, will foon fwell to feveral times the thicknefs it was before it was put in. When *Caffava* is intended to be laid up as a ftock to have recourfe to occafionally, or for the convenience of packing it up to fend about the country, it is then cured in the following manner: They put a parcel of the meal into a pan over a flow fire, and to prevent it from burning, or fticking to the pan, they continue ftirring it about with a wooden inftrument made for that purpofe. By this operation it is brought into granules, and when dry enough, it is taken out and laid by in fome convenient place, and by now and then expofing it to the fun, or the warmth of a ftove, it may be preferved fweet for feveral years. Whatever offal may happen to be made in any preparation of the root, is carefully faved, and dried in a ftove. This is often ufed to thicken their foups; but more generally, it is afterwards roafted very brown, and being fermented with the roafted roots of the *Convolvulus batatas* and melaffes, an inebriating liquor, called *ouycou*, is prepared from it, and is a favourite drink of the natives,

tives, and with which they moftly get intolerably intoxicated at their feafts and public
entertainments. No part of this extráor-
dinary root is wafted, for the juice, though
a perfect poifon crude, is boiled up with
meat, pepper, and other fpices, as occafion
requires, into a moft agreeable and wholefome foup; and they are very careful to
preferve it for this purpofe. Sometimes,
however, their hogs and poultry find means
to get at it, and drink it, which is inftant
death to them; yet the creatures fo poifoned, are eaten with the fame fafety and
unconcern, as if they had been properly
butchered.

Dr. Bancroft mentions another fort of
Caffava ufed by the Indians, which he calls
the fweet *Caffava*, and they *Camanioc*, and
fays it differs little from the former, but in
that it is not poifonous. This poffibly may
be the root of a fpecies of this genus, but it
certainly can never be a variety of the fame
plant. Notwithftanding its innocent quality,
its roots are not regarded by the natives as
equal to the others, they yielding lefs meal
in proportion to their fize, and that more
fpungy and lefs nutritive.

10 NYMPHÆA lotus. *Egyptian Lotus.*
Lin. Sp. pl. 729.
Nymphæa foliis amplioribus profundé
crenatis fubtus areolatis. *Brown. Jam.* 343.
This is an aquatick plant, and a native of
both

both the Indies. It fends up feveral large leaves, ftanding fingly on long footftalks; thefe are heart-fhaped, deeply cut at their bafe, of a light green colour, and fharply dentated on their edges. The flower-ftalks come immediately from the root; they are long, and each is terminated by one large, white double flower, of an agreeable fmell, and like that of our white Water Lily, but it is not quite fo full of petals. The calyx confifts of four permanent leaves, in the centre of which is placed the germen; this turns to a bottle-fhaped feed-veffel, of many cells, containing roundifh feeds.

The root is conical, firm, about the fize of a middling Pear, covered with a blackifh bark, and fet round with fibres. It has a fweetifh tafte, and when boiled or roafted, becomes as yellow within as the yolk of an Egg. The plant grows in abundance on the banks of the Nile, and is there much fought after by the poor people, who in a fhort fpace of time collect enough to fupply their families with food for feveral days.

11 SAGITTARIA fagittifolia. *Common Arrowhead. Lin. Sp. pl.* 1410.

Sagitta aquatica minor latifolia. *Bauh. Pin.* 194.

This plant grows common in rivulets and water ditches, and often varies much in the fize and form of its leaves. Ofbeck, in his Voyage to China, fays he faw *Sagittaria bulbis*

bulbis oblongis cultivated in the fame field
with *Rice* and *Nymphæa Nelumbo*; it re-
fembled the European *Sagittaria*, but was
larger, which might be owing to the culture: the roots of the Chinefe fort are the
fize of a clenched fift, and are oblong, and
the Swedifh are round, and not much larger
than peas. We change the quality of the
ground, 'he remarks, by draining the water,
and other arts, till we make it agreeable to
our few forts of corn; but the Chinefe make
ufe of fo many plants for their fubfiftence,
that they can fcarce have any fort of ground,
but what will fit fome one of them. Thus
they do not improve the field for the feed,
but chufe the feed for the field.

The *Sagittifolia* fends down into the mud
many long, flender, brittle fibres, with a
bulb fufpended at the end of each, which in
Auguft is about the fize of an Acorn, and of
a fine blue colour, ftreaked with yellow.
The infide is white, firm, of a farinaceous
tafte, but a little muddy. From the crown
of this bunch of fibres, fhoot many long,
fpungy ftalks, fupporting large arrow-fhaped leaves, of a fine green colour, and glofly
furface. Amidft thefe rife the flower-ftems,
higher than the leaves, fuftaining at their
joints three or four white flowers, on long
peduncles, each confifting of three roundifh
petals, which fpread open. The upper-
moft flowers are all male, with many awl-
fhaped

shaped stamina; the lower ones all female, with petals like the male, surrounding many compressed seed-buds, collected in a head, having very short styles, with acute stigmata. These flowers are succeeded by rough heads, containing many small seeds.

I cured some of the bulbs of this plant, in the same manner that Saloop is cured, when they acquired a sort of pellucidness; and on boiling them afterwards they broke into a glutinous meal, and tasted like old peas boiled.

12 Solanum tuberosum. *Common Potatoes. Lin. Sp. pl. 265.*

Solanum tuberosum esculentum. *Baub. Pin.* 167.

The common *Potatoe* is a native of Peru, in South America. It has been introduced into England about a century and half, but was amongst us a long time before much attention was paid to it, nor did it come into use in the families of the higher class of people, till within a few years past. The Irish seem to have been the first general cultivators of it in the western parts of Europe, and it is so extended now as to form a principal part of the winter food, both of the Irish and English. There are two sorts, the red and the white roots, which are only seminal variations; and there are also several varieties of these. *Potatoes* abound with an

3   insipid,

infipid, phlegmatic juice, which induces
many to think they are not nutritious; and
indeed fuch forts as break into a watery
meal in the boiling, can afford but very little
nourifhment, as they are always found to
prove very diuretick, and greatly to increafe the quantity of urine.   On the contrary, thofe kinds which cut firm when
thoroughly boiled, efpecially the white
forts, muft be nutritive, as they contain a
more mucilaginous juice, than thofe that
eafily break, which thickening in the boiling, is the occafion of the parts cohering.
Of equal quantities of the powder of Potatoes and the flour of Wheat, a good fort of
bread may be made; and ftarch and hair
powder may alfo be obtained from thefe
roots.

13 YUCCA gloriofa *.   *Adam's Needle.*
Lin. Sp. pl. 456.
Cordyline foliis pungentibus integerri-
mis. *Roy. Lugd. Bat.* 22.

This is a native of the fame place as the
former.   There are feveral fpecies of the
genus, all natives of America, but moft of
them are to be met with in the gardens and
green-houfes of the curious in England.
The *Gloriofa* differs from the reft, in having

* The plant that flowered at Coftefey, near Norwich, in
1732, and which was affirmed to be the *Succotrine Aloe,* was
only one of this fpecies; but it was a very ftrong plant, and
the ftem rofe to above fix feet high.

the

the margins of its leaves entire. In old plants the leaves are about eighteen inches long, and two broad, of a dark green colour, and each ends in a sharp stiff spine. They are thickly set round the bottom of the stem to a span or more upward, whence issues a round, rigid, purplish-green stalk, to the height of three feet or more, and which is set round with branches to the very top. At the base of each branch stands a small red leaf, with a green apex. The branches are sparedly set with bell-shaped flowers, which hang downwards; they are white, with purplish stripes on the outside, and consist of six petals each, joined together by their bases. In the center of the flower are six short, reflexed stamina, and an oblong, three cornered germen, which becomes an angular capsule, of three cells, filled with compressed seeds.

The root is thick and tuberous, and is used by the Indians for bread, being first reduced into a coarse meal; but this is only in times of scarcity, and when more grateful roots fail them. In like cases the people of England have been glad to support life with the roots of the *Spiræa filipendula*, (Dropwort) the *Scirpus maritimus*, (Baftard Cyperus) and even with those of the *Triticum repens*, (Dogs-grass) and also of those of the Common Brake, or Fern.

C   14 POLY-

14 POLYGONUM divaricatum. *Eastern Buckwheat. Lin. pl.* 520.

Perficaria alpina, folio nigricante, floribus albis. *All. Pedem.* 41. *t.* 8.

This grows in Siberia and the Iſland of Corſica, in the Mediterranean. 'Tis a perennial plant, with a creeping root, compoſed of many tough fibres. The ſtalk riſes near half a yard high, breaking into many ſpreading branches, which are moſtly bent at their joints, and are furniſhed with narrow, ſmooth, light green, ſpear-ſhaped leaves, ending in an acute point. The flowers are produced in looſe ſpikes at the ends of the branches; they have no calyx, are ſmall and white, conſiſt of one petal each, cut at the brim into five ſpreading ſegments, and contain eight ſtamina and three ſtyles. When the flower fades the petal enwraps a roundiſh, ſharp-pointed ſeed.

The roots (reduced into coarſe meal) are the ordinary food of the Siberians, as they are alſo of the mountain-rats. Theſe animals are provident enough in the winter to lay up a proper ſtore for the ſummer, which being known to the natives, and they being too indolent to dig for them, ramble in queſt of the rats, granaries, and having hit upon them, make no ſcruple to carry away the produce of all their induſtry.

SECT.

S E C T.   II.

*Roots occasionally eaten as Condiments, or for
other Family Purposes.*

1 AMOMUM zingiber. *Common Ginger.*
2 Allium cepa. *Common Onion.*
3 Allium afcalonicum. *Shallot, or Efchalot.*
4 Allium fcorodoprafum. *Rokambole.*
5 Apium petrofelinum. *Common Parfley.*
———— *latifolium.* Large-rooted Parfley.
6 Bunium bulbocaftanum. *Earth-nut, or
Pig-nut.*
7 Beta rubra. *Red Beet.*
8 Braflica rapa. *Common Turnep.*
———— *rapa punicea.* Purple - rooted
Turnep.
———— *rapa flavefcens.* Yellow-rooted
Turnep.
———— *rapa oblonga.* Long-rooted Turnep.
9 Campanula rapunculus. *Rampion.*
10 Cochlearia armoracia. *Horfe Radifh.*
11 Carum carui. *Caraway.*
12 Cyperus efculentus. *Rufh-nut.*
13 Daucus carota. *Wild Carrot.*
14 Eryngium maritimum. *Sea Holly.*

15 Guilandina moringa. *Ceylon Guilandina.*

16 Helianthus tuberofus. *Jerufalem Artichoke.*

17 Ixia chinenfis. *Spotted Ixia.*

18 Ixia crocata. *Greater African Ixia.*

19 Ixia bulbifera. *Bulb-bearing Ixia.*

20 Lathyrus tuberofus. *Peas Earth-nut.*

21 Orobus tuberofus. *Heath Peas.*

22 Orchis mafcula. *Male Orchis.*

23 Paftinaca fativa. *The Parfnep.*

24 Raphanus fativus. *The Radifh.*

25 Scorzonera hifpanica. *Viper's Grafs.*

26 Sium Sifarum. *Skirrets.*

27 { Lilium martagon. *Martagon Lily.*
   { Tulipa gefneriana. *Common Tulip.*

28 Tragopogon pratenfe. *Yellow Goatsbeard.*

29 Tragopogon porrifolium. *Purple Goatsbeard.*

1 AMOMUM zingiber. *Common Ginger. Lin. Sp. pl.* 1. Zingiber. *Bauh. Pin.* 35.

This is a native of both the Indies, and furnifhes a confiderable article of trade to the inhabitants of each. It is a perennial, and the roots fpread in the ground in digitated clufters. From thefe rife feveral reed-like ftalks, near a yard high, having a few narrow, graffy leaves towards their tops. Among thefe come forth the flower-ftems; they are naked all the way up, and terminated

nated by fcaly, oval fpikes of fmall blue
flowers, confifting of one irregular petal,
having a fhort tube; this is cut into four
fegments at the brim, and includes one ftamen and one ftyle. The germen becomes a
three-cornered capfule, containing many
feeds.

Ginger is an excellent ftomachick, and a
powerful expeller of flatulencies. The green
frefh root preferved as a fweetmeat, is preferable to any other. The Indians flice the
green root among their fallad herbs, in order to render them more grateful to the palate, and make them fit eafier on the ftomach.

2 Allium cepa. *Common Onion. Lin.
Sp. pl.* 431. Cepa vulgaris. *Bauh. Pin.* 71.

From whence this was firft brought into
Europe is not known, but that it is natural
to Africa is beyond a doubt, it being evident that *Onions* were eaten by the Egyptians above two thoufand years before
Chrift, and they make a great part of their
conftant food to this day in Egypt. Dr.
Haffelquift fays it is not to be wondered at
that the Ifraelites * fhould long for them,
after they had left this place, for whoever
has tafted *Onions* in Egypt muft allow, that
none can be had better in any part of the

* Numbers, chap. xi. ver. 5.

univerfe:

univerfe : here, he goes on, they are fweet, in other countries they are naufeous and ftrong; here they are foft, whereas in the north and other parts they are hard, and their coats fo compact, that they are diffi-cult to digeft. They eat them roafted, cut into four pieces, with fome bits of roafted meat, which the Turks call *kebab*; and with this difh they are fo delighted, that they wifh to enjoy it in Paradife. They likewife make a foup of them in Egypt, which Haffelquift fays is one of the beft difhes he ever eat. The many ways of dreffing *Onions* in England are known to every family, but in regard to wholefome-nefs, there is certainly no method equal to boiling, as thus they are rendered mild, of eafy digeftion, and go off without leaving thofe heats in the ftomach and bowels, which they are apt to do any other way. Their nature is to attenuate thick, vifcid juices, confequently a plentiful ufe of them in cold phlegmatick conftitutions muft prove beneficial. Many people fhun them on account of the ftrong, difagreeable fmell they communicate to the breath; this may be remedied by eating a few raw Parfley leaves immediately after, which will ef-fectually overcome the fcent of the *Onions*, and caufe them to fit more eafy on the fto-mach.

3 ALLIUM

3 ALLIUM afcalonicum. *Efchalot*. *Lin.*
Sp. pl. 429. Cepa fterilis. *Bauh. Pin.* 72.

This was found wild in `Paleftine, by
Dr. Haffelquift.   The root is conglobate,
confifting of many oblong roots, bound together by thin membranes. Each of thefe
fmall roots fends forth two or three fiftulous, long, awl-fhaped leaves, iffuing from
a fheath, and are nearly like thofe of the
common onion. The flower-ftem fhoots
from a membranaceous fheath, is round, almoft naked, and terminated by a globular
umbel of flowers, which have erect, purplifh, lance-fhaped petals, of the length of
the ftamina.

The root of this fpecies is very pungent,
has a ftrong, but not unpleafant fmell, and
therefore is generally preferred to the Onion,
for making high-flavoured foups and gravies.   It is alfo put into pickles, and in the
Eaft-Indies they ufe an abundance of it for
this purpofe.

4 ALLIUM fcorodoprafum.   *Rokambole*.
Lin. Sp. pl. 425.

Allium ftaminibus alternè trifidis, capite
bulbifero, fcapo ante maturitatem contorto.
Hall. all. 2.

This grows naturally in Denmark and
Sweden. It hath a heart-fhaped, folid root,
which ftands fide-ways of the ftalk. The
leaves are broad, and are a little crenated on

C 4   their

their edges. The flowers are of a pale purple colour, and collected into a globular head.

Linnæus makes the *Rokambole*, defcribed above by Haller, to be only a variety of this, and it differs from the original, in having the top of the ftalk twifted circularly before the flowers open, and alfo in the head producing bulbs. The roots are ufed for the fame purpofes as thofe of the former.

5 Apium petrofelinum. *Common Parfley.* *Lin. Sp. pl.* 379.

Apium hortenfe, petrofelinum vulgo. *Bauh. Pin.* 153.

The *Common Parfley* is known to every one. There are two varieties of it; the curled and the broad-leaved *Parfley*, the roots of which laft are frequently brought to the markets, efpecially the London ones. This variety has been cultivated in Holland a long time, and the roots are produced there to the fize of our fummer Carrots, which the gardeners tye up in bunches like Radifhes, and fend them to market, where they are readily bought by the people, who are very fond of them. They drefs them different ways, but the principal ufe they put them to is to make what they call Water-*Souché*. Parfley roots have a brifk diuretick quality, and therefore are not proper

2

food

food for fuch as have any debility in the urinary paffages. The plant is a native of the
Ifland of Sardinia.

6 Bunium bulbocaftanum. *Earth·Nut.*
Lin. Sp. pl. 349.
Bulbocaftanum majus, folio apii. *Bauh.
Pin.* 162.
This is a native of our woods and low
paftures. The leaves, as to their general
form, fomewhat refemble thofe of Parfley,
and thofe which come from the root lay flat
on the ground. The ftalk rifes to about
half a yard, is round, channelled, folid,
naked below, and divided upwards into
many branches, at each of which ftands a
fmall leaf, in fhape like thofe at the bottom. The flowers come out at the ends of
the branches in umbels; they are white,
and confift of five heart-fhaped petals each,
turning inwards, and furrounding five ftamina, with an oblong germen below, which
becomes an oval fruit containing two feeds.
The roots, which are of a dirty brown colour, and a little bigger than Hazel-nuts,
are as pleafant as a Chefnut, whence the
name of *Bulbocaftanum.* Pigs are exceedingly fond of thefe roots, therefore they are
called Pig-nuts; and indeed nature feems to
have intended them for the ufe of thefe
creatures rather than for man, by reafon they
cannot be improved by cultivation, as Potatoes

toes and other efculent roots are, for they will not thrive in tilled land. The root has a ftiptick quality, and has been deemed ferviceable againft laxity of the urinary paffages.

7 The *Beta rubra*, Red Beet, is only a variety of the *Beta vulgaris*, originally obtained by culture, and now there are fome varieties of this; as the common red Beet, the turneprooted red Beet, and the greenifh-leaved red Beet. This laft is the moft efteemed fort, the roots being larger and tenderer than the others. All thefe varieties are well known among gardeners, and the ufe of their roots among cooks; to defcribe them farther, therefore, would be ufelefs. They are pleafant enough to the palate, but are faid to be prejudicial to the ftomach, to afford little nourifhment, and on that account are but feldom eaten to what they were formerly.

8 BRASSICA rapa. *Common Turnep. Lin. Sp. pl.* 931. Rapa fativa rotunda. *Bauh. Pin.* 89. is a native of England, and may be met with wild on the borders of fields. No plant exhibits a more ftriking inftance of the benefits of cultivation than this, for in its wild ftate it is worth little either to man or beaft; but under the management of the hufbandman, it not only affords food for the human fpecies, but becomes a moft advantageous

vantageous crop to the cultivator, by fur-
nifhing the principal winter food for his
cattle. The Scotch eat the yellow-rooted
turneps, when fmall, as we do Radifhes;
and in France and Holland they boil the
long-rooted one in moft of their ftews and
gravies.

Turneps are an wholefome aperient food,
and the liquor preffed from them when
boiled is cooling and diuretick. The Turnep itfelf, mafhed with bread and milk,
is an excellent poultice.

9 CAMPANULA rapunculus. *Rampion.*
Lin. Sp. pl. 232. Rapunculus efculentus.
Bauh. Pin. 92.

The *Campanula rapunculus* grows wild in
the county of Surrey, and fome other parts
of England. It is a biennial plant with a
carrot-fhaped root, which fends forth many
elliptical leaves; among thefe rifes a firm,
erect, ftriated ftalk, to the height of two
feet, furnifhed with narrower leaves than
thofe from the root, ftanding irregularly.
Towards the top of the ftem, and at the bofoms of the leaves, rife feveral clofe panicles
of blue, bell-fhaped flowers, cut into five
fegments, and containing five ftamina and
one ftyle each. The whole plant abounds
with a lactefcent juice. It is much cultivated in France for the roots, which are
boiled

boiled and eaten as fallads; but in England it is now little regarded.

10 COCHLEARIA armoracia. *Horfe-radifh.* *Lin. Sp. pl.* 904. Raphanus rufticanus. *Bauh. Pin.* 97.

The root of the *Horfe-radifh* is perhaps one of the beft condiments to frefh beef, that the vegetable kingdom is capable of producing; for by its warmth and activity it promotes digeftion, and ftrengthens the tone of the ftomach. Frequently eaten, or otherwife ufed, it ftimulates the folids, attenuates the juices, fcours the glands, and thereby becomes ferviceable in fcurvies, and all diforders proceeding from a vifcid ftate of blood. The expreffed juice put into fkimmed milk makes an excellent cofmetic. There is a compound water of Horfe-radifh kept in the fhops, which is efteemed a good antifcorbutic. The plant grows naturally on the banks of rivers and ditches in England, and is too common to need a defcription.

11 CARUM carui. *Caraway. Lin. Sp. pl.* 378.

Carum pratenfe, Carui officinarum. *Bauh. Pin.* 158.

The *Caraway* is a biennial plant, and grows wild in our meadows and paftures.

It

It hath a carrot fhaped root, which runs deep in the ground, and which, on being broken, emits a ftrong aromatic fmell. From this comes up two or three folid, channelled ftalks, to about two feet high, fet with frefh green, winged leaves, on long footftalks, and more finely cut than thofe of the carrot. The ftalks break into branches upward, each of which is terminated by a bunch of fmall umbels, having white pentapetalous flowers, containing five hairy ftamina and one ftyle.

The roots of the cultivated *Caraways* were formerly in great efteem when boiled; how they have fallen into neglect is not eafy to guefs, as they certainly merit a place at table, as much as fome that come there, by reafon they have the faculty of warming and comforting a cold weak ftomach. The ufe of the feeds is well known both in the kitchen and fhops. There is an effential oil and fpirituous water drawn from the feeds, which are excellent Carminatives.

12 CYPERUS efculentus. *Rufh Nut. Lin. Sp. pl.* 67.

Cyperus rotundus efculentus anguftifolius. *Bauh. Pin.* 14.

This is a native of Italy, and a perennial. Immediately from the root fhoot up many long, narrow, graffy, three-fquare, fharp-pointed leaves, ftanding almoft upright; and
having

having a fharp, longitudinal ridge running
down the back of each. Amidft thefe rife
feveral, fmooth, three-fquare flower-ftems,
two or three feet high, each terminated by
five narrow leaves, fpreading horizontally,
from the centre of which comes an umbel
of flowers, compofed of four or five loofe
kind of panicles or rays, regularly difpofed,
bearing fmall, chaffy flowers, clofely crouded
together on each fide the midrib, and having three ftamina and one ftyle each.

The root is a collection of long fibres, fet
at fmall diftances with oval bulbs, which
are about the fize of nutmegs, of a reddifh
colour on the outfide, white within, firm,
and of a more delicate and pleafant tafte than
a chefnut. Thefe bulbs are greatly efteemed
in Italy and fome parts of Germany, and
are frequently brought to table by way of
defert.

13 DAUCUS carota. *The Carrot.* *Lin.*
Sp. pl. 348.
Paftinaca tenuifolia fylveftris Diofcoridis.
Bauh. Pin. 151.

The cultivated *Carrot* is well known to
every one, but there are many uninformed
of its being only a variety of the *daucus ca-*
rota, or wild carrot, fo common in our fields
and hedges. This, like the Turnep, is
worth little in its wild ftate, its root being
fmall, tough, and ftringy; yet when manured

nured it becomes large, fucculent, and of a
pleafant flavour. But even in its improved
ftate, unlefs eaten very young, it is hard of
digeftion, and confequently lies in the ftomach, and breeds flatulencies.

Both flowers and feeds of the wild Carrot
were kept in the fhops. The latter are a
powerful diuretick, and have often been
found a fovereign remedy in the jaundice,
dropfies and gravel.

14 ERYNGIUM maritimum. *Sea Holly.*
Lin. Sp. pl. 337.

This grows upon the fea coafts in diverfe
parts of England. It is a perennial, with a
long, tough, creeping root, which fends
forth feveral roundifh, plicated, bluifh,
prickly leaves, ftanding on long footftalks,
and moftly lodged on the ground. The
ftems rife about half a yard high, dividing
into many fpreading branches, which are
fet at their joints with leaves like thofe from
the root, but they are fmaller, and clafp
the ftalks with their bafe. The flowers are
produced at the ends of the branches in
roundifh, prickly heads, the bottoms of
which are furrounded with narrow, prickly
leaves, ranged in the form of a ftar. Each
flower confifts of five fmall, oblong, light-
blue petals, furrounding five flender ftamina and one ftyle. The germen becomes

an

an oval fruit, divided into two cells, each containing one oblong feed.

The roots have a pleafant, fweetifh tafte, mixed with a flight degree of warmth and acrimony. They are candied by the confectioners, and eaten in this manner they are deemed excellent for diforders of the breaft and lungs.

15 GUILANDINA moringa. *Ceylon Guilandina. Lin. Sp. pl.* 546.

Lignum peregrinum aquam cæruleam reddens. *Bauh. Pin.* 416.

This grows in Egypt, the Ifland of Ceylon, and on the coaft of Malabar. It is a fhrubby tree, and the only one of the genus that has no fpines; the others, four in number, being all armed with prickles. This rifes with a ftrong ftem, covered with an afh-coloured bark, to near twenty feet. The young branches are covered with a green bark, and fet at their bafe with trifoliate leaves, but upon the branches the leaves are decompounded, breaking into feveral divifions, which are again divided into fmaller ones, having five pair of oval lobes each, and terminated by an odd one. Thefe are of a light-green colour, and a little hoary on their under fide. The flowers are produced from the fides of the branches, in loofe bunches; they are yellow, compofed

of

of an unequal number of petals, some having five and others ten, and stand in a bell-
shaped calyx, which is cut at the brim into
five equal parts. The stamina are awl-
shaped, ten in number, and surround an
oblong germen, which becomes a rhomboidal pod, with one cell, including several
hard, oval seeds.

The root is thick, full of knobs, and
when young, is scraped and used by the inhabitants in the same manner, and for the
same purposes, as we do Horse-radish, it
having the like pungent taste, as have also
the flowers. The wood of this tree dyes a
beautiful blue colour.

16 HELIANTHUS tuberosus. *Jerusalem
Artichoke. Lin. Sp. pl.* 1277.

Helenium indicum tuberosum. *Bauh.
Pin.* 277.

The *Jerusalem Artichoke* is a native of
Brazil, but has for ages been cultivated in
the English gardens. It is a perennial, and
sends up many round, hairy, stiff stalks,
eight or ten feet high, which are set with
yellowish green, oval heart-shaped leaves,
somewhat like those of the common Sun-
flower, but narrower. A farther description
of it will be needless, it being pretty well
known among gardeners; for where it has
once been planted, it is no easy matter to
root it out again. The roots have some re-
D   semblance

femblance to Potatoes, but their tafte is more fulfome, and like that of Artichoke bottoms. They abound with a phlegmatic juice, which is apt to generate wind, and caufe uneafy griping pains in the bowels. This is the chief reafon they are not fo much cultivated now as they were formerly.

17 IXIA chinenfis. *Spotted Ixia. Lin. Sp. pl.* 52.

Bermudiana iridis folio majori, flore croceo eleganter punctato. *Krauf. hort.* 25. *t.* 25.

This is a perennial, and a native of India. It hath a thick, flefhy, jointed root, furnifhed with fibres. This fends up a fmooth, jointed ftalk, fet with pointed leaves, near a foot long, and an inch broad, with furrows running their whole length, and clafping the ftalk with their bafe. Some way up, the ftalk divides into two, and a peduncle fhoots from the centre of the partition, fupporting one flower; thefe two branches divaricate again into peduncles, about two inches long, each fuftaining a flower as the former. The flower confifts of fix equal petals, of a deep gold colour on the outfide, but of a light yellow within, mixed with red fpots; in the centre are three ftamina and one inclining ftyle. The germen is oval, three cornered, and ftands

below

below the flower; this turns to a capfule with three cells, filled with roundifh feeds.

The inhabitants where the plant grows naturally, boil the roots, and cut them as we do potatoes.

18 Ixia crocata. *Greater African Ixia.* *Lin. Sp. pl.* 52.

Ixia foliis gladiatis glabris, floribus co-rymbofis terminalibus. *Mill. ic.* 156. *f.* 1.

This hath a flattifh, bulbous root, fending forth three or four thin, narrow, fword-fhaped leaves, near a foot long, among which rifes the flower-ftem juft above them. The ftem is very flender, naked, and ter-minated by a fpike of yellow flowers, com-pofed of fix large, oblong, concave petals, of a glaffy hue at their bafe, where each has a large, blackifh fpot on the infide.

19 Ixia bulbifera. *Bulb-bearing Ixia.* *Lin. Sp. pl.* 51.

This from a bulbous root fends forth fe-veral narrow, fword-fhaped leaves, about half a foot long. Among thefe rifes a jointed ftem, to near half a yard, which is furnifhed with a fmall leaf at each of its lower joints, clafping the ftem with its bafe, and ftanding erect. At the bofoms of thefe leaves bulbs are produced, which if planted will vegetate, and produce com-pleat plants. The flowers come out alter-

D 2   nately

nately at the upper part of the stem, which
bends at the joints where they spring from;
they are composed of six whitish oval petals
each, striped with blue on their outsides.
The germen supports a long, slender style,
crowned with a trifid stigma, and turns to a
roundish capsule, having three cells, filled
with small roundish seeds.

Thefe two last species are natives of the
Cape of Good Hope, where the roots are
eaten by the inhabitants, and greatly
esteemed. There are several more of this
genus, and it is probable the roots of all
of them might be used in the same manner.

20 LATHYRUS tuberofus. *Peas Earth
Nut. Lin. Sp. pl.* 1033.

Lathyrus arvensis repens tuberofus. *Bauh.
Pin.* 344.

In the corn-fields of France and Germany
the *Peas Earth Nut* grows naturally, and is
a very troublesome weed to the farmers. It
is a perennial, and strikes some of its fibres
very deep into the earth, whilst others run
obliquely near the surface, having thick
knobs, or irregular bulbs at their ends.
From the crown of the bundle of fibres
come several trailing stalks, three or four
feet long, and furnished with oval, sessile
leaves in pairs, with a clasper between
them. The flowers are produced from the
arm-pits of the leaves, three or four upon a
long

long peduncle ; they are of the pea kind,
of a light purplish colour, and are succeeded
by slender, curved pods, containing small,
round seeds.

This plant, though a weed in France, is
cultivated in Holland for the roots, which
are carried to the markets there for sale.
They have an agreeable pleasant taste, much
resembling that of the Sweet Chesnut.

21 OROBUS tuberosus. *Heath Peas. Lin.
Sp. pl.* 1028.

Astragulus sylvaticus, foliis oblongis glabris. *Baub. Pin.* 351.

This grows plentifully on the heaths in
Scotland, and also on the like places in
some parts of the north of England. This
too is a perennial plant, having a more
woody root than the *Lathyrus* above-men-
tioned. It sends up a simple stem, about a
foot high, furnished with winged leaves,
generally composed of two pair of oblong-
oval, smooth, sharp-pointed lobes each, and
a sort of triangular stipula at the base of the
footstalk, which embraces the stem. From
the joints of the stem spring the peduncles,
each supporting three or four flowers of the
pea kind, which turn to a deep purple before
they fall.

The roots of this when boiled are said to
be nutritious. They are held in great
esteem by the Scotch Highlanders, who

chew them as we do Tobacco, and thus
often make a meal of them; for being of a
fedative nature, they pall the appetite, and
allay the fenfation of hunger, the fame as
Tobacco does.

22 ORCHIS mafcula. *Male Orchis. Lin.
Sp. pl.* 1333.
Orchis foliis feffilibus non maculatis.
Bauh. Pin. 82.
This is very common in our woods,
meadows, and paftures, and the powdered
roots of it are faid to be the Saloop, which
is fold in the fhops; but the fhop roots
come from Turkey. The flowers of moft
of the plants of this genus are indifcriminately called *Cuckoo-flowers* by the country
people. Though it has been affirmed that
Saloop is the roots of the *mafcula* only, yet
thofe of the *morio*, and of fome other fpecies
of *Orchis*, will do equally as well, as I can
affirm from my own experience; confequently to give a defcription of the *mafcula*
in particular will be ufelefs. As moft
country people are acquainted with thefe
plants, by the name of *Cuckoo-flowers*, it
certainly would be worth their while to
employ their children to collect the roots for
fale; and though they may not be quite fo
large as thofe that come from abroad, yet
they may be equally as good, and as they
are exceedingly plentiful, enough might annually

nually be gathered for our own confumption, and thus a new article of employment would be added to the poorer fort of people. The time for taking them up is when the feed is about ripe, as then the new bulbs are fully grown; and all the trouble of preparing them is, to put them fresh taken up into fcalding hot water for about half a minute; and on taking them out to rub off the outer fkin; which done, they muft be laid on tin plates, and fet in a pretty fierce oven for eight or ten minutes, according to the fize of the roots; after this, they fhould be removed to the top of the oven, and left there till they are dry enough to pound.

Saloop is a celebrated reftorative among the Turks, and with us it ftands recommended in confumptions, bilious cholics, and all diforders proceeding from an acrimony in the juices. Some people have a method of candying the roots, and thus prepared they are very pleafant, and may be eaten with good fuccefs againft coughs and inward forenefs.

23 PASTINACA fativa.   *The Parfnep.*
Lin. Sp. pl. 376.

Paftinaca fylveftris latifolia.   *Bauh. Pin.*
155.

The *Paftinaca* is found wild upon banks and the mere-balks of fields, and differs from the garden Parfnep only in the fize of its root, and the hairinefs of its leaves, the cultivated one having a larger and more flefhy root, and fmoother leaves. The roots of the garden Parfnep feem to claim the preference to all other efculent roots, of Englifh growth, they being very agreeable to moft palates, eafy of digeftion, and afford excellent nourifhment. In the northern parts of Ireland the poor people obtain a fort of beer from thefe roots, by mafhing and boiling them with hops, and then fermenting the liquor. The feeds of the wild Parfnep are flightly aromatic, and are often kept in the fhops.

24 RAPHANUS fativus. *The Radifh. Lin. Sp. pl.* 935.

Raphanus minor oblongus. *Bauh. Pin.* 96.

This was originally brought from China, and by cultivation there are now in the gardens here feveral varieties of it; for befides the long-rooted black Spanifh Radifh, we have two or more forts with round roots. Radifhes abound with almoft an infipid watery juice, which is apt to breed flatulencies. The outer fkin has a brifk pungency, and therefore fhould never be fcraped off, as this much corrects the phlegmatic part.

Radifhes boiled are fcarcely to be excelled by Afparagus. For this purpofe they ought to be rather fmall and frefh drawn, and then drefled in the fame manner that Afparagus

is.

is. They are a long time before they become tender; it moſtly taking an hour to
boil them ſufficiently.

25 SCORZONERA hiſpaniça. *Viper's Graſs.*
Lin. Sp. pl. 1112.

Scorzonera latifolia ſinuata. *Bauh. Pin.*
275.

Spain and Siberia are the native places of
the *Viper's Graſs.* It is a perennial, and hath
a tap-ſhaped root, about the thickneſs of one's
finger, blackiſh without, and white within,
of a bitteriſh ſub-acrid taſte, and abounds
with a milky juice, as does the whole plant.
The firſt leaves are large, ſinuated on their
edges, and end in a long acute point.
Among theſe riſes the ſtem to near three
feet. This is ſmooth, much branched towards the top, and irregularly ſet with long,
narrow leaves, whoſe baſe partly embrace it.
Each branch of the ſtem terminates with a
long, ſcaly head, compoſed of many narrow,
tongue-ſhaped, hermaphrodite florets, laying
over each other, and of a bright yellow
colour, ſomewhat reſembling the yellow
Goat's-beard. The florets are ſucceeded by
oblong, whitiſh, rough ſeeds, crowned with
feathery down.

The root is not only an article of cook-
ery, but alſo of confectionary, it being
preſerved with ſugar in the manner of
Eryngo.

It

It was formerly a celebrated alexiphar-
mick, and in great esteem for strengthening
the stomach, and promoting the fluid secre-
tions. The juice too has been deemed a
counter poison to the bite of the Viper,
hence the plant obtained the name of *Viper's
Grafs*.

26 SĪUM Sifarum. *Skirrets. Lin. Sp. pl.*
361.

Sifarum germanorum. *Bauh. Pin.* 155.

This is a native of China, but has been a
long time cultivated in most parts of Europe, and particularly in Germany. The
root is a bunch of fleshy fibres, each of
which is about as thick as a finger, but
very uneven, covered with a whitish, rough
bark, and has a hard core or pith running
through the centre. From the crown of
this bunch come several winged leaves, con-
fisting of two or three pair of oblong, dentated
lobes each, and terminated by an odd one.
The stalk rises to about two feet, is set with
leaves at the joints, and breaks into branches towards the top, each terminating with
an umbel of small white flowers, which are
fucceeded by striated feeds like those of
Parsley.

Skirrets come the nearest to Parsneps of
any of the esculent roots, both for flavour
and their nutritive quality. They are rather sweeter than the Parsnep, and there-

2   fore

fore to some few palates are not altogether
so agreeable.

It is evident from experiments which have
been made on this and some other vegetables, that bounteous nature has not confined sugar to the Indies only, but has liber-
ally blended it in the constitution of many
European plants, and which may, by proper management, be extracted from them of
equal quality, and perhaps nearly as copiously as from the celebrated Sugar-cane. The
ingenious Chymist, M. Margraaf, has given
some experiments he made on the roots of
the *Beet* and *Skirret*, in order to obtain this
valuable commodity from them; and as he
found the latter to yield it in the greatest
quantity, and by reason too it is a matter
both curious and important, I shall here
give his process in as concise a manner as
the subject will admit.

He took a quantity of fresh *Skirret-roots*
well cleaned, and having cut them into
small pieces, beat them to a mash in an iron
mortar; then tying them up in a linen bag,
he committed them to a press, and squeez-
ed them till the juice would run no longer. Water was then poured upon the same
mashed roots, and they were put into the
press in a bag the second time, and pressed
as before: the liquor obtained by these two
operations was kept in a cool place for forty
eight hours, when it became clear, and had
precipi-

precipitated a mealy fubftance to the bottom of the veffel in which it was contained. Finding the fæces thus fettled, he poured the clear liquor through a fine linen cloth into a frefh veffel. To this ftrained liquor he added fome whites of eggs, and then boiled the whole together in a copper pan, frequently fkimming it, till no fæces appeared on the furface, but the liquor became as tranfparent as the pureft clarified wine. It was then again boiled in a fmaller pan, till a confiderable part was evaporated; and the fame operation was continued till the original thin liquor was become of the confiftence of common fyrup. The boiling being compleated, he fet the thickened liquor in a warm place for fix months, at the end of which time the fugar was fhot in the form of cryftals about the fides of the veffel.

To feparate and purify this fugar was the next and main operation, and in order to this he immerfed the veffel in warm water, thereby to break the tenacity of the liquor, and render it more fluid. This done, he pours the whole into an earthen pot, having a wide mouth, and narrow bottom pierced with holes, and placing this within another pot, fet both of them in a temperate warmth for fome time. By this contrivance, the liquid part fell gradually through the perforations of the firft pot, into the fecond, and left the cryftals remaining in the firft.

This

This fugar was coarfe and clammy, and therefore to bring it more pure, he wrapped it up in a piece of blotting paper, and then gently preffed it with his hand; the effect was, that the paper fucked up much of the vifcid moifture that had adhered to the fugar, and thereby left the latter more neat. Having thus divefted it of its groffeft impurities, he again boiled it up with lime-water till it became ropy, and taking it off the fire, kept ftirring it till near cold, when he poured it off into a conical earthen veffel, ftopped with wood. This he placed in another veffel as before, and in the fpace of about eight days, the fyrup had all dropped through the firft veffel, and left the cryftals behind. Thefe he purified ftill farther by means of blotting paper, as before, and a parcel of neat fugar was procured, equal in goodnefs to the beft produced from the Sugar-cane. The liquid that was faved in the laft pot too, had all the properties of common melaffes.

It muft be confeffed that this procefs of Margraaf's, to extract the fugar from plants, is both flow and tedious; but neverthelefs, it points out how copioufly fome of our vegetables are ftored with a faccharine falt, which might be drawn from them in abundance by proper management, or an eftablifhed method of bufinefs, as they have for the Sugar-cane; and that if it fhould ever happen,

happen, that we were entirely deprived of this valuable article from abroad, yet the means of furnishing ourselves with it exists in our own country. By a shorter, but more expensive process, the same gentleman extracted sugar from several other roots, as *Carrots, Parsneps,* &c. and from the *Beet* and *Skirret* he has set down the qualities as follow : from

½ ℔. of White Beet root, ½ oz. of pure Sugar, ½ ℔. of Red Beet root, 1¼ oz. of pure Sugar, ½ ℔. of Skirret root, 1½ oz. of pure Sugar.

Of these, he says, the sugar obtained from the White Beet was the best, that from the Skirret was next in goodness, and consequently the Red Beet afforded the worst of all.

$$27 \begin{cases} \text{LILIUM martagon.} & Lin.\ Sp.\ pl. \\ \quad 435. \\ \text{TULIPA gesneriana.} & Lin.\ Sp.\ pl. \\ \quad 438. \end{cases}$$

I have placed the *Lilium martagon* and the *Tulipa gesneriana* together, for the convenience of speaking upon them under one head. The first is a native of Hungary and some places of Siberia, and the latter grows spontaneously in Asia Minor. Linnæus says that the roots of the *Martagon Lily* make part of their daily food in Siberia, and that those of the *Tulip* are eaten in several parts

of

of Italy. This may feem ftrange to thofe
who never had heard of fuch an ufe being
made of them, but there are feveral other
roots which were formerly made ufe of in
diet, that are now totally neglected. Some
fpecies of *Ornithogalum* furnifhed a conftant
difh for the poorer people, where the plants
grew fpontaneoufly, and the root of the *latifolium* in particular was deemed excellent.
I prefume a great many bulbous roots of
plants in the Hexandria Clafs, might be introduced into diet with fafety and advantage; efpecially fuch as have little fmell
and tafte, for that great mafter of nature,
Profeffor Linnæus, has laid it down as a
rule, that fuch plants as are no ways offenfive to the palate and organs of fmell, are of
themfelves of a harmlefs nature. And on
the contrary, that thofe that are immediately difgufting to thefe two faculties, ought
to be rejected as hurtful and pernicious.
The firft part of this rule is confirmed by
daily experience, for all the forts of grain
conftantly introduced into human food,
have nothing in them, even in their crude
ftate, that is obnoxious to either of thefe
fenfes. And in refpect to roots, we find
nothing difagreeable in the flavour or fmell
of raw *Turneps*, *Parfneps*, *Potatoes*, and
others, and when dreffed they prove both
pleafant and nutritious. As to the fecond
part of the rule, I conceive Linnæus's meaning

ing to be this; that fuch plants as affect
the organs with a very uneafy fenfation, are
improper for conftant food; for if he intended they muft not be eaten at all, experience fhews the contrary. *Onions, Garlick,*
and many more, whofe fmell is difagreeable
to fome, are occafionally ufed in diet, and
in a general way are found to be wholefome.
The roots of the *Crown Imperial* have a
very naufeous fmell, yet are frequently ftewed in foups, without yielding any noxious
quality to the liquor perceiveable in the
quantity ufed; but this does not by any
means prove, that they may be generally
eaten with fafety.

28 Tragopogon pratenfe. *Yellow Goatsbeard. Lin. Sp. pl.* 1109.

Tragopogon pratenfe luteum majus.
Bauh. Pin. 274.

This is a biennial plant, and grows very
common on the borders, and mere-balks of
our corn-fields. It hath a tap-fhaped root,
which fends forth a few narrow, graffy
leaves, ending in an acute point, and doub-
led, fo as to make their edges nearly meet.
The ftalk rifes more than half a yard high,
fet at its joints with leaves like thofe at the
bottom, and embracing the ftalk with their
bafe. Sometimes, near the top, the ftalk
breaks into two or three branches, each being terminated with a long, green, conical

2   bud

bud, which on its breaking spreads horizontally, and displays numerous yellow,
tongue-shaped, hermaphrodite florets, cut
into five teeth at their points, and laying
over each other like tiles. These are nearly
equal in length to the rays of the empalement, and are succeeded by oblong, pointed seeds, crowned with long, feathery down,
the whole forming a regular globe of two
or three inches diameter.

The plant is known by the country people under the name of *Go to bed at noon*, or
Sleep at noon, it being peculiar to the flowers to close themselves in the middle of the
day. They dig up the roots when young
and dress them as Parsneps, to which they
prefer them.

29 *Tragopogon* porrifolium. *Purple Goatsbeard. Lin. Sp. pl.* 1110.

Tragopogon purpuro-cæruleum, porrifo-
lio, quod artefi vulgo. *Bauh. Pin.* 274.

This too is a biennial, and is found wild
in Cornwall, and some other places in England. It is a much larger plant than the
former, and has leaves somewhat resembling those of Leeks; but they are of a
lighter green colour, and each has a white
longitudinal line running through its centre.
The stalk is terminated with a large, beautiful purple flower, having the rays of the
empalement much longer than the florets;

E   and

and juft below the flower, it fwells fo as to become thicker than in the other parts.

This plant is cultivated in gardens by the name of *Salfafy*, and its roots are dreffed and ferved up at table in a variety of forms. They are of a pleafant, nutritious nature, but though thefe are at prefent in the greateft efteem, they are much inferior to thofe of the *pratenfe*.

C H A P.

ESCULENT SHOOTS, STALKS, SPROUTS
AND PITHS.

S E C T. I.

Firſt Shoots and Stalks.

1 **A**SPARAGUS officinalis. *Aſpa-ragus.*
2 Anethum azoricum. *Sweet Azorian Fennel.*
3 Angelica archangelica. *Angelica.*
4 Arctium lappa. *Common Burdock.*
5 Aſclepias Syriaca. *Greater Syrian Dogſ-bane.*
6 Apium graveolens. *Smallage.*
——— dulce. *Garden Celery.*
7 Campanula pentagonia. *Thracian Bell-flower.*
8 Cynara cardunculus. *Cardoon, or Char-doon.*
9 Carduus marianus. *Milk Thiſtle.*
10 Cnicus cernuus. *Siberian nodding Cni-cus.*
11 Chenopodium bonus Henricus. *Engliſh Mercury.*

E 2   12 Con-

12 Convolvulus foldanella. *Sea Bindweed.*

13 Cucubalus behen. *Spatling Poppy.*

14 Epilobium anguftifolium. *Rofebay Willow-herb.*

15 Humulus lupulus. *Wild Hops.*

16 Onopordum acanthium. *Cotton Thiftle.*

17 Rheum rhaponticum. *Rhapontick Rhubarb.*

18 Smyrnium olufatrum. *Common Alexanders.*

19 Smyrnium perfoliatum. *Round-leaved Alexanders.*

20 Saccharum officinarum. *Sugar-cane.*

21 Sonchus alpinus. *Mountain Sow-thiftle.*

22 Tamus communis. *Black Briony.*

23 { Tragopogon pratenfe. *Yellow Goatsbeard.*
Tragopogon porrifolium. *Purple Goats-beard.*

1 ASPARAGUS officinalis. *Lin. Sp. pl.* 448.

Afparagus maritimus, craffiore folio. *Bauh. Pin.* 490.

The wild *Afparagus* differs little from the garden, except in the finenefs of the leaves. The latter is fo generally cultivated as to require no defcription, and the agreeablenefs of its young fhoots, as a fallad, need not be mentioned. They certainly promote the appetite, but are faid to afford little nourifhment. By the ftrong, fœtid fmell they

they communicate to the urine, foon after eaten, it is evident they are diuretick; but the plant in its wild ftate is faid to be more powerfully fo, than in its manured one. It is a native of England, and grows in the marfhes near Briftol.

2 ANETHUM azoricum. *Sweet Azorian Fennel*.

Though this is made a diftinct fpecies of *Fennel* by fome writers on Botany, yet it certainly is no other than a variety of the *Anethum fœniculum* of Linnæus, which is the common *Fennel*. It was originally brought from the Azorian Iflands, in the Atlantick ocean, hence the trivial name *azoricum*. The plant is much cultivated by the Italians, under the name *Finochio*. It is low, compared with the common *Fennel*, and differs from it too in the nature of its ftalks; which, inftead of running up, begin to fpread as foon as they get above the furface of the ground, till they become four or five inches broad, very flefhy, and fome-times near two inches thick.

The ftalks have a fweet, fulfome tafte, mixed with an aromatic, and are eaten ei-ther raw with oil and vinegar, or ftewed in foups and gravies.

3 ANGELICA archangelica. *Angelica. Lin. Sp. pl.* 360.

Angelica fativa. *Bauh. Pin.* 155.

E 3 Lapland

Lapland is the native country of this plant, where it grows in great plenty upon the banks of the rivers.

The root confifts of a parcel of thick flefhy fibres, fending forth feveral large, compound winged leaves, of a lightifh green colour, having broad flefhy footftalks, and are compofed of oblong, ferrated, fharp-pointed lobes. Among thefe rifes a round, fiftulous, jointed ftalk, to the height of five or fix feet, and fet with leaves at the joints, whofe membranous bafes embrace it. Towards the top the ftem breaks into many branches, each terminated by a compound umbel, the rays of which are angular, and fupport globular heads of whitifh flowers, containing five ftamina and one ftyle each. Thefe are fucceeded by greenifh feeds, ftanding by pairs.

The ftalks were formerly blanched and eaten as Celery, and the young fhoots are at prefent in great efteem among the Laplanders. The plant is one of the fineft aromatics Europe produces. Gardeners who have water running through their grounds cultivate it for the roots, which they fell to the confectioners to be made into a fweet-meat. This confection is one of the moft warm and agreeable that can be; is good to expel wind and ftrengthen the ftomach, and is furpaffed only by that of Ginger.

4 ARCTIUM lappa. *Common Burdock.*
Lin. Sp. pl. 1143.

Lappa major five arctium diofcoridis.
Bauh. Pin. 198.

The *Arctium lappa* is a biennial plant,
and is very common in wafte grounds and
by road fides.

It hath a long, thick, brown root, fending
out many exceeding large heart - fhaped,
greyifh-green leaves; among which rifeth a
purplifh, tough, ftriated ftalk, divided into
feveral branches, furnifhed with fmaller
leaves. At the extremities of the branches
come the flowers in bunches; they confift
of a multitude of purple, hermaphrodite
fiorets, included in a fcaly empalement,
thickly fet with long, flender, incurved
fpines.

Many people cut the tender ftalks of this
plant, and having ftripped off the outer
fkin, boil and drefs them like Afparagus.
They have not a very pleafant flavour, but
the plant being aperient and fudorific, a
frequent eating them in this manner would
certainly do good fervice in fcorbutic habits.
A decoction of the roots has been found to
be very beneficial againft the rheumatifm,
gout, and other diforders bordering upon
thefe. I myfelf have lately been a witnefs to
their good effects this way. If the boiled
ftalks, or a decoction of the roots, fhould be
difagreeable to any, who may be defirous

E 4   of

of using them for the above complaints, they may preserve either with sugar, and eat them as a sweetmeat, but they will not prove altogether so efficacious.

5. Asclepias Syriaca. *Greater Syrian Dogsbane. Lin. Sp. pl. 313.*

Apocynum majus syriacum rectum. *Corn. canad. 90.*

The *Asclepias Syriaca* is a native of Virginia, but has been a long time planted in the English gardens, both on account of its being an exotic, and for the sweet smell of its flowers, which are nearly as fragrant as those of the *Hesperis tristis*, or Garden Sweet Rocket.

From a white creeping root it sends up many round stems, four or five feet high, at the joints whereof stand two sessile, bright green oval leaves, opposite each other.

At the tops of the stalks, and sometimes at the bosoms of the leaves, come forth almost globular umbels of small, yet low purplish flowers, consisting of one petal each, divided into five oval parts, and containing five very minute stamina and one style. In the centre are two oval germina, which become two oblong, pointed pods, filled with compressed seeds, crowned with a soft down. The whole plant is so full of a milky juice, that when a leaf is taken off, the wound will discharge for a considerable time after.

This

This plant has been always deemed a fatal poifon to dogs, and very dangerous to the human fpecies; notwithftanding this, a Mr. Wagftaff, of Norwich, has lately made trial of its young fhoots, by boiling and dreffing them like Afparagus, and they proved equally as pleafant and well tafted. Nor did the eaters experience any bad effects from them; hence it may be concluded, that either the boiling deftroyed their deleterious property, or that the young fhoots did not poffefs it in a hurtful degree. That fire will deftroy the pernicious qualities of plants, is evident from the management of the *Jatropha maniot* before mentioned, and is alfo farther evinced by the leaves of Tea, which are poifonous in their crude ftate, but by being dried over an oven, this quality is fo diminifhed or blunted, as not to be fenfibly felt in a moderate ufe of them. It may reafonably be concluded then that boiling had a great fhare in rendering this plant falubrious, and that many others which are now deemed hurtful, might thereby become wholefome and agreeable fallads.

As the *Afclepias* was found to be innocent and palatable in almoft a natural ftate, if it were put under the art and management of the gardener, much might be expected from it; for it being a plant that will thrive in any foil and fituation, and as it fends out a prodigious quantity of fuckers from the

root,

root, its propagation would be eafy, and in time it might be made to anfwer all the purpofes of Afparagus, and prove a good fuccedaneum to that celebrated, but expenfive vegetable. The down or cotton that adheres to the feeds of this plant, and fome others of the fame genus, and which is called *delawad* in France, is there made ufe of for ftuffing of chairs and quilts. The latter are extremely warm and light, and are exceeding proper coverings for people labouring under any infirmity of body; for this matter is fo elaftic, that it adds little to the weight. A French gentleman has lately hit upon a method of fpinning this down into balls like filk, for which contrivance he has obtained a patent from the French council, authorizing him to fabricate it into velvets and other ftuffs.

6 Apium graveolens. *Smallage.* *Lin. Sp. pl.* 379.

Apium paluftre five Apium officinarum. *Bauh. Pin.* 154.

The *Apium graveolens* grows upon the banks of moift ditches in England, and fometimes even in the water. It is cultivated in gardens by the name *Apium dulce,* or Celery. In its wild ftate it is faid to have very noxious qualities, but by cultivation it becomes not only wholefome, but ferviceable for ftrengthening the ftomach

and

and affifting digeftion. Moft umbelliferous plants that grow in the water, or moift places, are poifonous, or at leaft hurtful to the human frame; but by tranfplanting they lofe their evil qualities, and become aromatic and carminative. Celery is now fo generally known as to render a defcription of the plant ufelefs; nor need it be mentioned, that the blanched ftalks are eaten raw, ftewed, or otherwife.

7. CAMPANULA pentagonia. *Thracian Bell-flower.* *Lin. Sp. pl.* 239.

Speculum veneris, flore ampliffimo, Thracicum. *Raj. Hift.* 742.

This grows in Thrace, and alfo in the corn-fields in France. It is a low, annual plant, feldom rifing more than feven or eight inches. The ftalks are numerous, weak, very much branched, and near their bottom have five obfolete angles.

The leaves are linear, that is, almoft all the way of a breadth, fharpifh pointed, and have no footftalks.

The flowers come out both at the divifions of the ftalks, and the extremities of the branches; they are of a bluifh purple, with a white eye in the centre; are deeply cut at their brims into five round fegments, and contain five fhort ftamina and one club-fhaped ftyle each. The feed veffel is long, triangular,

triangular, deeply furrowed, and contains many compreſſed, brown ſeeds.

The firſt tender ſhoots of this plant are a favourite ſallad among the French. They ſow it thick, and cut it when ſmall as we do creſſes. It has an agreeable taſte, ſomewhat like Corn Sallad, and is held to be a good antiſcorbutic. It is known in our gardens by the name of *Thracian Venus Looking - glaſs*.

8 CYNARA cardunculus. *Cardoon.* Lin. Sp. pl. 1159.

Cinara ſpinoſa cujus pediculi eſitantur. *Bauh. Pin.* 383.

The *Cynara cardunculus* is a native of Candia, formerly the Iſland of Crete, in the Mediterranean ſea. It differs from the common Artichoke in growing much taller, in the leaves being more finely cut, and thicker ſet with ſpines, and in having ſmaller and rounder heads.

The gardeners blanch the ſtalks, as they do Celery, and they are eaten raw with oil, pepper and vinegar; or as fancy directs they are boiled or ſtewed, and ſometimes laid upon a toaſt and cheeſe.

9 CARDUUS marianus. *Milk Thiſtle.* Lin. Sp. pl. 1153.

Carduus albis maculis notatus vulgaris. *Bauh. Pin.* 281.

This

This is plentiful in waste places, and upon old banks. It is known to almost every one by its large, beautiful leaves, which are variegated with white spots and veins, as if they had been sprinkled with milk. This circumstance gave rise to a foolish, monkish tradition, that the Virgin Mary, when suckling our Saviour, accidentally let fall her milk upon the leaves of this plant, which stained all the succeeding ones since. The young shoot for leaves in the spring, cut close to the root, with part of the stalk on, is one of the best boiling sallads that is eaten, and surpasses the finest Cabbage.

10 CNICUS cernuus. *Nodding Cnicus.*
Lin. Sp. pl. 1157.

Siberia is the native country of the *Cnicus cernuus.* It is a perennial plant, with a thick, fleshy root, that breaks into many turgid fibres.

The radical leaves are heart-shaped, near a foot long, and six or seven inches broad; they stand upon very short footstalks, are of a deep green colour on their upper side, whitish underneath, and sawed on their edges.

The stem is reddish, generally near six feet high, channelled, and furnished with leaves more heart-shaped than those at the root. Towards the top it divides into

branches,

branches, each terminated by a globular head of yellowiſh flowers, ſurrounded by a ſcaly, prickly empalement. The florets are all hermaphrodite, funnel-ſhaped, cut into five ſegments at their brims, and contain five ſhort, hairy ſtamina and one ſtyle each.

The tender ſtalks are firſt peeled, and then boiled and eaten as a ſallad, by the inhabitants where the plant grows.

11 CHENOPODIUM bonus Henricus. *Engliſh Mercury. Lin. Sp. pl. 318.*
Lapathum unctuoſum. *Bauh. Pin. 115.*

The *Engliſh Mercury* is frequently to be met with in waſte, and rubbiſhy places. From the root, which conſiſts of ſeveral thick fibres, come forth many arrow-ſhaped, dark green leaves. Among theſe riſe the flower-ſtalk, to about eighteen inches, thickly crouded with leaves, and divided at the top into many greeniſh ſpikes of flowers, having no petals, but conſiſt of a penta-phyllous * calyx each, containing five ſtamina and one ſtyle.

The young ſhoots boiled are by many eſteemed beyond Spinach, and it was formerly cultivated in the Engliſh gardens the ſame as Spinach now is, but of late it has been neglected, though it certainly merits the attention of the gardener as much as any ſallad in preſent uſe.

* Compoſed of five leaves.

The

The country people call the plant *All-good*, from a conceit that it will cure all hurts; and the leaves are now a conſtant plaiſter among them for green wounds.

12 CONVOLVULUS ſoldánella.   *Sea Bind-weed. Lin. Sp. pl.* 226.
Soldanella maritima minor. *Bauh. Pin.* 295.

The *Convolvulus ſoldanella* is common upon our ſea-coaſts, where the inhabitants gather the young ſhoots, and pickle them in the manner of Samphire. They have a cathartic quality, for a ſmall quantity of the pickle will gently move the bowels. They have a ſalt, bitteriſh taſte before pickled.

The plant hath a ſlender, creeping root, which puts forth many weak, purpliſh, ſtriated ſtalks, about half a yard long, and trail upon the ground; theſe are furniſhed with kidney-ſhaped leaves, ſupported on long footſtalks, and ſomewhat reſembling thoſe of Pilewort. The flowers come out at the footſtalks of the leaves, on long pe-duncles; they are of a deep red colour, and bell-ſhaped, like thoſe of common Bind-weed. The whole plant abounds with a milky juice.

13 CUCUBALUS behen.   *Spatling Poppy. Lin. Sp. pl.* 591.

Lychnis

Lychnis fylveftris, quæ Behen album vulgo. *Bauh. Pin.* 205.

This is perennial, and very common in corn-fields and hedges. It hath a whitifh, creeping root, compofed of many joints, whence fpring feveral ftalks, about half a yard or two feet long, having their bottom parts curved, and ufually lodge upon the ground; thefe are very full of joints near their bafe, and thickly fet with pea-green, lance-fhaped leaves, ftanding oppofite, embracing the ftalks with their bafe. The lower leaves are moftly finely ciliated on their margins. The flowers come out plentifully at the tops of the ftalks; they are compofed of five white bifid petals, protruded from a bladdery calyx, with a ftamen inferted in the tail of each petal, and five ftanding alternately between them, the number of ftamina being ten. The ftyles are uncertain, fome flowers having but three, others four, and fome five.

Our kitchen-gardens fcarcely furnifh a better flavoured fallad than the young, tender fhoots of this plant, when boiled. They ought to be gathered upon tilled land, and when they are not above two inches long. If the plant were under cultivation, no doubt but it would be improved, and would well reward the gardeners labour, by reafon it fends forth a vaft quantity of fprouts, which might be nipped off when of a proper fize, and

and there would be a fucceffion of frefh ones for at leaft two months. It being a perennial too, the roots might be tranf-planted into beds, like thofe of Afpara-gus.

The dried roots were formerly kept in the fhops, by the name of *Behen album*, and were deemed cordial and cephalic.

14 EPILOBIUM anguftifolium. *Rofebay Willow-herb. Lin. Sp. pl.* 493.

Lyfimachia chamænerion dicta angufti-folia. *Bauh. Pin.* 245.

This is a perennial plant, and common in woods and meadows in the northern parts of England. The radical leaves rife in a tuft; they are long, narrow, fharp-pointed, of a deep glofly green on the upper furface, of a filvery grey underneath, entire at their margins, have no footftalks, and have fe-veral tranfverfe veins running through their fubftance. In the centre of thefe rifes a round, firm, upright ftem, to a man's height, irregularly fet with leaves like the former, to near the top, where the ftem is terminated by a long racemus of large, beautiful, deep-red flowers, ftanding in quadrifid calyces, and compofed of four roundifh petals each, furrounding eight de-clining ftamina and one ftyle. The germen is cylindrical, placed below the flower, and

F   turns

turns to a capfule of five cells, filled with
oblong feeds, crowned with down.

The young tender fhoots cut in the
fpring, and dreffed as Afparagus, are little
inferior.

15 HUMULUS lupulus. *Wild Hops. Lin.
Sp. pl.* 1457.

Lupulus mas et femina. *Bauh. Pin.* 298.

This is the only fpecies of the genus, and
is to be found wild in our-hedges. It is
male and female in diftinct plants, and is fo
well known by being generally cultivated,
as to render a defcription of it ufelefs. The
young fhoots are often gathered by the poor
people, and boiled as an efculent fallad. If
they be chofen very young they are good
and pleafant; but if too far advanced, they
are then tough, bitter, and ftringy.

In regard to the medicinal virtues of the
flowers of this plant, they are one of the
moft agreeable and ftrongeft bitters. Their
principal ufe is in malt liquors, which they
render lefs glutinous, and difpofe them to
pafs off more freely by urine.

16 ONOPORDUM acanthium. *Cotton
Thiftle. Lin. Sp. pl.* 1158.

Spina alba tomentofa latifolia fylveftris.
Bauh. Pin. 382.

This is a biennial plant, and is to be
found plentifully in uncultivated places in
many

many parts of England. The root is long and fibrous, and fends forth feveral oblong, fharp-pointed, whitifh green, finuated leaves, covered with a cottony down; and fet with fpines on their edges. In the midft of thefe fhoots up a ftalk, to the height of five or fix feet, divided towards the top into diverfe branches, fet with leaves at their joints, and having jagged, leafy borders running along them, edged with fpines, as has the main ftalk alfo. Each branch terminates with a fcaly head of reddifh purple, hermaphrodite florets, having narrow tubes, and cut at their brim into five teeth. They contain five hairy ftamina and one ftyle, and are fucceeded by fmall oblong feeds, crowned with down.

The tender ftalks of this plant, peeled and then boiled, are greatly efteemed by many, whilft the fingular flavour they have is difagreeable to fome few palates.

17 RHEUM rhaponticum. *Rhapontic Rhubarb. Lin. Sp. pl.* 531.

Raponticum folio lapathi majoris glabro. *Bauh. Pin.* 116.

This is an inhabitant of the mountain Rhodope, in Thrace, but has been a long time cultivated in the Englifh gardens. It is a large, perennial plant, with a thick, flefhy root, which divides into many parts as big as Parfneps, running deep in the

ground.

ground.   It is of a reddifh brown colour on
the outfide, yellow within, and fends forth
many very large, fmooth, heart-fhaped leaves,
having thick footftalks of a reddifh green
colour, which are a little channelled on
their under fide, but are flat on the upper.
When the plant grows in rich, ftrong land,
the leaves will be two feet long, and as much
broad, and they will have large, prominent
veins running from the infertion of the footftalk to the borders.   The footftalk too will
be as long as the leaves, and thicker than a
man's finger at their bafe.   The leaves are
of a dark green colour, flightly waved on
their edges, and have a fubaftringent tafte,
mixed with an acid.   Among thefe leaves
rifes the flower-ftem, to the height of two
or three feet; this is of a purplifh colour,
mixed with green, and has at each joint a
fmall feffile leaf, of the fhape of the former.
The flowers are produced at the top of the
ftalk, in clofe, obtufe panicles; they are
very fmall, have no empalements, but confift of one whitifh petal each, cut into fix
fegments, and having nine flender ftamina
inferted into it, furrounding three fhort,
reflexed ftyles.   The feeds are large, brown,
triangular, and winged.

The footftalks of the radical leaves having
an acid tafte, and being thick and flefhy,
are frequently ufed in the fpring for making
of tarts.   If they be carefully peeled they
will

will bake very tender, and eat agreeably. Many people prefer them even to Apples. There is another fpecies of this genus *(the compactum)*, the ftalks of which I have many times known to have been ufed in the fame manner, and have been counted equally as good; and I am inclined to think that the ftalks of all the fpecies might be thus employed indifcriminately.

The *Rhaponticum* was introduced into Europe in the beginning of the feventeenth century, by Alpinus, and was then fup-pofed to be the true Rhubarb. The root is undoubtedly the Rhubarb of the ancients, but it is far inferior to either of the forts kept at prefent in the fhops, it being but flightly cathartic, and much more aftringent. A decoction made from the green frefh roots is an excellent antifcorbutic, and in this refpect is no way excelled, if equalled, by a decoction of the fo much celebrated Wa-ter-Dock.

18 SMYRNIUM olufatrum. *Common Alex-anders. Lin. Sp. pl.* 376.

Hippofelinum theophrafti, five Smyrnium diofcoridis. *Bauh. Pin.* 154.

Since the introduction of Celery into the garden, the *Alexanders* is almoft forgotten. It was formerly cultivated for fallading, and the young fhoots or ftalks blanched were eaten either raw or ftewed. The leaves too

were

were boiled in broths and foups. It is a warm comfortable plant to a cold, weak ftomach, and was in much efteem among the monks, as may be inferred by its ftill being found in great plenty by old abbey walls.

It is a biennial, and hath a long, white root, which fends forth winged leaves, fomewhat like thofe of Smallage, but much larger, and the lobes rounder. The ftalk is furrowed, rifes four or five feet high, is divided into many branches, and furnifhed with leaves at the joints. The branches terminate with large umbels of greenifh white flowers, having five fmall, inflexed, fpear-fhaped petals each, including five ftamina of the fame length, and two ftyles. The natural foil of this plant is on rocks near the fea, and it is found in fuch places in the north of England and Scotland.

19 SMYRNIUM perfoliatum. *Round-Leaved Alexanders.* *Lin Sp. pl.* 376.

Smyrnium peregrinum rotundo, five oblongo folio. *Bauh. Pin.* 154.

The *Perfoliatum* is a native of Italy. The bottom leaves of this fpecies are exceedingly beautiful, being decompounded of many frefh green, fmall leaves, which are divided into three oval, ferrated lobes each. The ftalk rifes in the centre of thefe firft leaves to about three feet high, and is divided near the

the top into two or three branches. It is fmooth, hollow, and jointed. At each joint ftands one large orbicular leaf, of a yellowifh green colour, plain on the margin, and clafps the ftalk with its bafe. This change of the leaves, from compound wing-ed ones, to thofe that are round and entire, gives the plant a very fingular appearance. The branches are terminated by compound umbels of fmall, yellowifh flowers, having the fame number of petals and ftamina as thofe of the *olufatrum*.

The blanched ftalks of this fpecies are far preferable to thofe of the former, they being more pleafant and much tenderer.

20 SACCHARUM officinarum. *Sugar-cane. Lin. Sp. pl.* 79.

Arundo faccharifera. *Bauh. Pin.* 18.

This plant grows naturally in both In-dies, where it is alfo cultivated for that ufe-ful part, its juice. It has a jointed root, which fends forth feveral fhoots, that arrive to a heighth according to the nature of the foil. The medium one, however, is nine or ten feet. Thefe ftalks are jointed, and each joint has a leaf two or three feet long, which embraces the ftem with its bafe to the next joint above it, before it expands. The ftalks are of a light yellow colour, of a brittle fubftance, and have a white fweet pith running through them. The leaves

F 4 are

are narrow, sharp pointed, set with fine
sharp teeth on their edges, like those of the
Schœnus mariscus, and have a whitish prominent rib running from their apex to their
base.   The flowers are produced at the tops
of the stalks, in large panicles, in the manner of our common Reed; these have no
calyx, but each is composed of a bivalved,
acute-pointed glume, surrounded with long,
woolly down, and contains three hair-like
stamina, of the length of the glume, together with an awl-shaped germen, supporting two rough styles, crowned with simple
stigmata.   The germen becomes an oblong,
acute-pointed seed, enclosed in the glume.

The young, tender shoots are boiled, by
the inhabitants of the West-India Islands,
with Bananas, and Spanish Potatoes, into a
thick pottage, there called *Callulos*.   This is
for negroe food, and is both pleasant to the
palate, and very nourishing.   The shoots
thus boiled too, are exceedingly agreeable,
if eaten by themselves.

Nature scarcely produces a more valuable
plant than the *Sugar-cane*; for though it is
not immediately necessary to the support of
human life, yet it is capable of adding greatly to its comforts and enjoyments.   Beside
furnishing us with several home made wines,
it would be impossible to reap the benefit of
many sorts of fruit, in the manner we do, if
we were entirely deprived of the sweet, delicious

licious falt, called Sugar. By the mollify-
ing qualities of this, many acid fruits are
rendered palatable and agreeable in pies,
tarts, &c. By this, feveral kinds of berries
and roots are preferved from putrefaction
from year to year, and become ufeful both
in food and medicine. Rum, which is
made from the produce of the Sugar-cane,
is an excellent oily, nourifhing fpirit, if
ufed phyfically, and in proper quantities.
This, Melaffes, and Sugar, furnifh a prodigious fund of trade and riches, both to the
inhabitants of the Indies, and thofe of Europe. To lay before the reader the tedious
procefs of extracting the Sugar from the
Canes, would be only abufing his time, as
this has been fully treated upon by feveral
writers, and it may be fuppofed he is already
acquainted with it; I muft therefore farther
regard the immediate ufefulnefs of the plant,
and obferve, that in the Indies, the tops of
the *Canes* are cut into fmall pieces, and giv-
en to their domeftic cattle, to which they
prove very nourifhing food, and keep them
fat and in good fpirits.

21 SONCHUS alpinus. *Mountain Sow-
thiftle. Lin. Sp. pl.* 1117.
Sonchus lævis laciniatus, five Sonchus alpinus cæruleus. *Bauh. Pin.* 124.
This *Sonchus* is a native of England, and
is found on the fides of hills. It is common

2   too

too in Lapland, where the inhabitants eat the young shoots as a sallad. How they may suit an English palate I don't know, but those who have a mind to try may obferve the following defcription of the plant.

It is an annual, and fends up a ftraight, round, hollow, purplifh ftem, irregularly fet with jagged leaves, fomewhat like thofe of Dandelion, but the finufes are finely ferrated on the edges. The flowers come out at the top of the ftem in a racemus; they are large, and compofed of many blue, hermaphrodite florets, ftanding in an imbricated, bellying calyx. The feeds are like thofe of the common Sow-thiftle, crowned with down.

22 TAMUS communis.   *Black Briony.*
Lin. Sp. pl. 1458.

Bryonia lævis five nigra racemofa. *Bauh.*
Pin. 297.

The *Black Briony* is common in woods and hedges in moft parts of England. It is male and female in diftinct plants. The root of either is large, tap-fhaped, flefhy, and covered with a dark brown fkin. From this come feveral brownifh green ftalks, which twine about any thing within their reach, till they arrive at ten or twelve feet in length; thefe are furnifhed at the joints with dark green, gloffy, heart-fhaped leaves, ftanding fingly upon long foot-ftalks.

The

The flowers are produced in short, turgid bunches from the sides of the stalks; those of the male have six short stamina, fixed to a flat empalement, of six oval leaves; the females are composed of a bell-shaped calyx or empalement, cut into six segments, with an oblong, punctured gland sitting on the inside of each. When the female flowers are fallen, they are succeeded by smooth, dark red berries, of the size of small Grapes, containing six round seeds each, about as big as those of Gromwell.

This plant has been generally held to have corrosive and dangerous qualities, yet its young shoots are frequently boiled and eaten in the spring, the same as those of Hops, and are by many as much esteemed. The leaves and roots were formerly kept in the shops; the latter, scraped and then rubbed upon any part pained or swelled with the rheumatism, has in most instances done much service. When thus used they ought to be fresh taken out of the ground.

23 Tragopogon pratense. *Yellow Goatsbeard.*

Tragopogon porrifolium. *Purple Goatsbeard.*

Both these were described in the former Chapter, therefore it is only to be observed here, that their young shoots, when advanced to about four inches high, are

boiled

boiled and eaten in the manner of the reft of this order, and that thofe of the *pratenfe* are frequently preferred to Afparagus. Both plants contain a milky juice, have a diuretic quality, and are fuppofed ferviceable againft the gravel.

SECT. II.

Sprouts and Piths.

1 ARECA oleracea. *Cabbage-tree.*
2 Arundo bambos. *Bamboo-cane.*
3 Braffica oleracea, vel fylveftris. *Sea, or Common White Cabbage.*
———— *viridis.* Green Savoy Cabbage.
———— *fabauda.* White Savoy Cabbage.
4 Braffica botrytis. *Cauliflower.*
———— *alba.* White Cauliflower Brocoli.
———— *nigra.* Black Cauliflower Brocoli.
5 Braffica fabellica. *Siberian Brocoli, or Scotch Kale.*
6 Braffica præcox. *Early Batterfea Cabbage.*
7 Braffica rapa. *Common Turnep.*
8 Cyperus papyrus. *Paper Rufh.*
9 Cycas circinalis. *Sago Palm-tree.*
10 Portulaca oleracea. *Purflane.*
———— *latifolia.* Broad-leaved Garden Purflane.

11 Smi-

11 Smilax afpera. *Red-berried rough Bind-*
weed.

1 ARECA oleracea. *Cabbage-tree.* *Lin.*
Syftema Naturæ, 730.

This is a fpecies of Palm, and a native of
the Weft-Indies, where it grows with a
taper body to a very great height. The
leaves are large and pinnated, and the lobes
are entire. It hath male and female flowers iffuing from the fame fpatha. The male
are fupported upon a branched fpadix,
fpringing from a bivalve fpatha; thefe have
three fharp pointed, ftiff petals each, furrounding nine ftamina, three of which are
longer than the reft. The female flowers
come from the fame common fpatha, have
no ftyles, but confift of three acute-pointed
petals each. When thefe fall off, the germen fwells to an almoft oval, fibrous berry,
containing one oval feed.

This is the only fpecies of Palm that is
mentioned by Linnæus to afford efculent
leaves or buds, and it is from the pith of
this fpecies that the Weft-India Sago is faid
to be made; but whether this is the only
one that bears what is called the *Cabbage,*
is not eafily to be determined, by reafon the
defcriptions given by different writers of
this kind of tree, are very vague and uncertain. Miller, in his Dictionary, has mentioned two forts of efculent Palms, one
from

from Sloane, which he calls the *Cabbage-tree*, and the other from Dr. Houstoun, which he names the *Mountain Cabbage*. To what particular genus of the Palms either species belongs, is impossible to be told from Miller; and by the small difference in the descriptions * given by the above two gentlemen of the trees, it is probable they both meant one and the same species, and that the *Areca oleracea* of Linnæus. But as this is a matter which cannot be made perfectly clear, I shall describe Miller's Cabbage - trees in his own words. "This tree rises to a very great height in the country where it grows naturally. *Ligon*, in his history of *Barbadoes*, says, there were then some of the trees growing there, which were more than two hundred feet high; and that he was informed they were a hundred years growing to maturity, so as to produce seed. The stems of these trees are seldom larger than a man's thigh; they are smoother than those of most other sorts, for the leaves naturally fall off entire from them, and only leave the vestigia or marks where they have grown. The leaves (or branches) are twelve or fourteen feet long; the small leaves or lobes are about a foot

---

* Palma altissima non spinosa, fructu pruniformi minore racemoso sparso. *Sloan. Cat.*
Palma altissima non spinosa, fructu oblongo. *Houstoun.*

long,

long, and half an inch broad, with feveral
longitudinal plaits or furrows, ending in
foft acute points, and are placed alternately.
The flowers come out in long loofe bunches
below the leaves; thefe branch out into
many loofe ftrings, and are near four feet
long, upon which the flowers are thinly
placed. The female flowers are fucceeded
by fruit about the fize of a Hazel-nut, having a yellowifh fkin, fitting clofe to the
ftrings of the principal footftalk.

"As the inner leaves of this encompafs
the future buds more remarkably than moft
of the other fpecies, fo it is diftinguifhed by
the appellation of Cabbage-tree; for the
centre fhoots, before they are expofed to the
air, are white and very tender, like moft
other plants that are blanched; and this is
the part which is cut out and eaten by the
inhabitants, and is frequently pickled and
fent to England by the title of Cabbage;
but whenever thefe fhoots are cut out, the
plants decay and never thrive after; fo that
it deftroys the plants, which is the reafon
that few of the trees are now to be found in
any of the Iflands near fettlements, and
thofe are left for ornament."

This is Miller's account of Sloane's Cabbage tree; and of that defcribed by Houftoun,
he fays, "The fruit of this kind is about
an inch and an half in length, and near two
inches in circumference. The flower-buds,
which

which are produced in the centre of the plants, are by the natives cut, and boiled to eat with their meat, and are by them called the Mountain Cabbage."

From thefe accounts of the two trees we find, that the buds for leaves of the one are eaten, and the flower-buds of the other, which feem to indicate, indeed, that they are diftinct fpecies; unlefs it may be, that both forts of buds of the fame tree are ufed as here mentioned. In regard to the genus *Areca*, it contains only two fpecies at prefent known, the *oleracea*, and the *cathecu*; from the juice of the latter the *Terra Japonica* of the fhops is faid to be made. This laft tree is called *Faufel*. Captain Dampier, in his voyage round the world, met with abundance of *Cabbage-trees* at the Ifland of St. Jago, near the Ifthmus of Darien, in the South Sea, where he meafured one which reached 120 feet; the leaves or branches were 12 or 14 feet. In the middle of the branches, he fays, grows the fruit, (this is, the Cabbage) about a foot long, as thick as a man's leg, white, and very fweet, whether eaten raw or boiled. Between the Cabbage and the branches fprout many fmall twigs, about a foot long, at the end of which grow hard, round berries, of the fize of Cherries, which falling off, afford excellent food for the hogs.

In the body of thefe trees there runs a
pith,

pith, which is a nutritious food to a particular kind of worms, and in which, after
the trees are felled, they breed in great
abundance.   These worms or grubs are
eaten by the French in some of the West
India islands, and are esteemed a great delicacy.   They are nearly the size of one's
little finger, and have a black head, equal
in thickness to the body.   The manner of
dressing them for table is, to string them
upon skewers, and hang them before the
fire, and as soon as they are thoroughly
warm, to sprinkle them with fine raspings
of crust of bread, salt, pepper, and nutmeg,
thereby to absorb the fat.   When sufficiently
roasted they are served up with orange or
citron sauce.

2 ARUNDO bambos. *Bamboo Cane. Lin.
Sp. pl.* 120.
Tabaxir & Mombu arbor.   *Bauh. Hist.*
I. *p.* 222.
This curious *Reed* is a native of both the
Indies, where it frequently attains the height
of sixty feet.   The main root is long, thick,
jointed, spreads horizontally, and sends out
many cylindrical, woody fibres, of a whitish
colour, and many feet long.   From the
joints of the main root spring several round,
jointed stalks, to a prodigious height, and
at about ten or twelve feet from the ground,
send out at their joints several stalks joined
G   together

together at their bafe; thefe run up in the fame manner as thofe they fhoot out from. If any of thefe be planted, with a piece of the firft ftalk adhering to them, they will perpetuate their fpecies. They are armed at their joints with one or two fharp, rigid fpines, and furnifhed with oblong - oval leaves, eight or nine inches long, feated on fhort footftalks. The flowers are produced in large panicles, from the joints of the ftalks, placed three in a parcel clofe to their receptacles; they refemble thofe of the common Reed, and are fucceeded by feeds of the fame form, furrounded with down.

The young fhoots are covered with a dark green bark; thefe when very tender are put up in vinegar, falt, garlic, and the pods of capficum, and thus afford a pickle, which is efteemed a valuable condiment in the In- dies, and is faid greatly to promote the ap- petite, and affift digeftion. The ftalks in their young ftate are almoft folid, and con- tain a milky juice; this is of a fweet nature, and as the ftalks advance in age, they be- come hollow, except at the joints, where they are ftopped by a woody membrane, upon which this liquor lodges, and con- cretes into a fubftance called *Tabaxir*, or fugar of *Mombu*, which was held in fuch efteem by the ancients, in fome particular diforders, that it was equal in value to its weight in filver. The old ftalks grow to

5   five

five or fix inches diameter, are then of a
fhining yellow colour, and are fo hard and
durable, that they are ufed in buildings.
Thefe, when bored through the membranes
at their joints, are converted into water-
pipes, and make excellent good ones.   The
fmaller ftalks are ufed for walking-fticks,
and are called *Rotang*.   The inhabitants of
Otaheite make flutes of them, about a foot
long, with two holes only, which they ftop
with the firft finger of the left hand, and
the middle one of the right, and they blow
through their noftrils.

3 Brassica oleracea.   *Common White
Cabbage*.   *Lin. Sp. pl.* 932.

This is a native of England, and is found
wild on the fea-coafts.   Numbers 4, 5, and
6, are, by Linnæus, made only varieties of
Number 3.   Whether he is right in this is
hard to determine, for the number of *Cabbages* now raifed makes it impoffible to tell
with certainty, which are fpecies and which
varieties.   And this difficulty is conftantly
increafing by the mixing of the farina of
one fort with another, and thereby producing new variations.   There is fome probability, however, that the Cauliflower is a
diftinct fpecies, and it is certain that the
different forts of Brocoli are varieties of the
Cauliflower.   They are all too well known
to require any defcription, and their young

G 2   fhoots

shoots are generally acknowledged to be superior to most other vegetables.

7 BRASSICA rapa. *Common Turnep. Lin.
Sp. pl.* 931.
Rapa sativa rotunda. *Bauh. Pin.* 89.
This has been mentioned in the first
Chap. but as the sprouts are frequently eaten
in the spring, it had a right to a place here
also. If these be gathered when very tender, they are an excellent sallad.

8 CYPERUS papyrus. *Paper Rush. Lin.
Sp. pl.* 70.
Papyrus Syriaca et Siciliana. *Bauh.
Pin.* 19.
This is a grafs-leaved plant, growing
naturally in Egypt, Syria, and some other
parts of the East. It hath a creeping root,
from which comes forth a tuffuck of long,
slender leaves; in the midst of these rise
very thick three-square, naked stalks, terminated by umbels of small, chaffy flowers,
laying over each other like tiles. The
spokes or rays of the umbel are long, slender,
exceedingly numerous, stand rather upright,
and are nearly of an equal length. Each
iffues from a short distinct sheath, and towards the top is set with awl-shaped, sessile
spiculæ, standing by threes on a short peduncle. The flower contains three short
stamina, tipped with oblong summits, and

5   one

one flender ftyle, fupporting three hair-like ftigmata. The germen is fmall, and becomes a three-fquare, fharp-pointed feed.

The ftalk of this plant contains a fweet, nutritious pith, which the ancient Egyptians made ufe of as bread. Of the ftalks or leaves, it is now uncertain which, they made their paper, but the manner of preparing it is at prefent unknown. It feems, however, to have been the only paper in ufe in the time of Mofes. The Egyptians likewife made fails and even boats of thefe rufhes, which they caulked with flime and pitch, and in one of thefe Mofes was concealed by his mother *.

9 CYCAS circinalis. *Sago Tree.* Lin. *Sp. pl.* 1658.

Palma indica, caudice in annulos protuberante diftincto. *Raii Hift.* 1360.

This is a fpecies of Palm, which grows fpontaneoufly in the Eaft Indies, and particularly on the coaft of Malabar. It runs up with a ftraight trunk, to forty feet or more, having many circles the whole length, occafioned by the old leaves falling off; for they ftanding in a circular order round the ftem, and embracing it with their bafe, whenever they drop, they leave the marks

---

* Exodus Chap. ii. ver. 3. Thefe boats are ftill in ufe in the eaftern parts of Africa, where they are kept upon the lakes as pleafure-boats.

of

of their adhefion behind. The leaves are
pinnated, and grow to the length of feven
or eight feet. The pinnæ or lobes are long,
narrow, entire, of a fhining green, all the
way of a breadth, lance-fhaped at the point,
are clofely crouded together, and ftand at
right angles on each fide the midrib, like
the teeth of a comb. The flowers are produced in long bunches at the footftalks of
the leaves, and are fucceeded by oval fruit,
about the fize of large plums, of a red colour when ripe, and a fweet flavour. Each
contains a hard brown nut, enclofing a white
meat, which taftes like a Chefnut.

This is a valuable tree to the inhabitants
of India, as it not only furnifhes a confiderable part of their conftant bread, but
alfo fupplies them with a large article of
trade. The body contains a farinaceous
fubftance, which they extract from it and
make into bread in this manner: they faw
the body into fmall pieces, and after beating them in a mortar, pour water upon the
mafs; this is left for fome hours to fettle.
When fit, it is ftrained through a cloth,
and the finer particles of the mealy fubftance running through with the water, the
grofs ones are left behind, and thrown away.
After the farinaceous part is fufficiently fub-
fided, the water is poured off, and the meal
being properly dried, is occafionally made
into cakes and baked. Thefe cakes are faid
to

to eat nearly as well as wheaten bread, and are the fupport of the inhabitants for three or four months in the year.

The fame meal more finely pulverized, and reduced into granules, is what is called *Sago,* which is fent into all parts of Europe, and fold in the fhops for a great ftrengthener and reftorative.

There is a fort of Sago made in the Weft Indies, and is fent to Europe in the fame manner as that from the Eaft; but the Weft India Sago is far inferior in quality to the other.   It is fuppofed to be made from the pith of the *Areca oleracea,* already defcribed.

10 PORTULACA oleracea.   *Purflane.* *Lin. Sp. pl.* 638.

Portulaca anguftifolia fylveftris.   *Bauh.* *Pin.* 288.

This is an annual plant, and a native of the warmer parts of Europe.   It has many round, thick, reddifh, fucculent ftalks, near half a yard long, which generally lodge upon the ground; thefe break into many branches, thickly fet with feffile, flefhy, wedge-fhaped leaves, fome of a pale green, others of a reddifh colour, and moftly ftanding four or more together in whorls.   In the bofoms of thefe the flowers are produced; they are feffile, fmall, of a yellowifh colour, and are compofed of five plain erect, obtufe petals each, with many hair-like ftamina, about

G 4   half

half the length of the petals, and one ſtyle, crowned with oblong ſtigmata.

This plant is frequently raiſed in gardens for ſallading, and the alteration it receives from culture is chiefly in the breadth and ſucculency of the leaves. Many admire it, but it is of a very cold nature, and apt to chill the blood, therefore ſhould be eaten ſparingly.

11 SMILAX aſpera. *Red-berried rough Bindweed. Lin. Sp. pl.* 1458.

Smilax aſpera, fructu rubente. *Bauh. Pin.* 296.

This ſpecies of *Smilax* is a ſhrubby plant, and grows ſpontaneouſly in Spain, Italy, and Paleſtine. It hath a fleſhy root, which ſends up ſeveral weak, brown, ſlender, angular ſtalks, armed with ſhort, crooked ſpines, and are furniſhed with tendrils at their joints or bents, by which they claſp round any adjacent plant, and by that means riſe to ſeven or eight feet high. The leaves are large, ſtiff, heart-ſhaped, very ſharp-pointed, of a reddiſh colour, have ſhort reddiſh ſpines on their margins, and are ſupported on ſlender footſtalks. The flowers are produced in ſmall bunches at the angles of the ſtalks, and are male and female on ſeparate plants. The male flowers are compoſed of a ſix-leaved bellying calyx, containing ſix ſtamina, crowned by oblong ſummits. The
femalеs

females too have no petals, each confifting
of a calyx like the male, with an oval germen, fupporting three ftyles. The fruit is
a fmall red berry, having three cells, containing two feeds each.

The young tender fhoots are boiled and
eaten as others of this order. There are two
or three varieties of this plant, one in particular with a black fruit.

C H A P.

C H A P. III.

ESCULENT LEAVES.

SECT. I.

Cold Sallads.

1 APIUM petrofelinum. *Parſley.*
————— crifpum. *Curled-leaved Parſley.*
2 Allium cepa. *Common Onion.*
3 Allium fchœnoprafum. *Cives.*
4 Allium oleraceum. *Wild Garlick.*
5 Artemifia dracunculus. *Taragon.*
6 Alfine media. *Common Chickweed.*
7 Borago officinalis. *Borage.*
8 Cacalia ficoides. *Fig Marigold - leaved Cacalia.*
9 Cichorium endivia. *Endive.*
————— endiva crifpa. *Curled-leaved Endive.*
10 Cochlearia officinalis. *Scurvygraſs.*
11 Eryfimum alliaria. *Jack by the Hedge.*
12 Eryfimum barbarea. *Winter Creſs or Rocket.*
13 Fucus faccharinus. *Sweet Fucus or Sea Belts.*
14 Fucus palmatus. *Handed Fucus.*

15 Fucus

15 Fucus digitatus. *Fingered Fucus.*
16 Fucus efculentus. *Edible Fucus.*
17 Hypochæris maculata. *Spotted Hawk-weed.*
18 Lactuca fativa. *Lettuce.*
19 Leontodon taraxacum. *Dandelion.*
20 Lepidium fativum. *Garden Crefs.*
21 Lepidium virginicum. *Virginian Sciatic Crefs.*
22 Mentha fativa. *Marfh or Curled Mint.*
23 Mentha viridis. *Spear Mint.*
24 Oxalis acetofella. *Wood Sorrel.*
25 Poterium fanguiforba. *Garden Burnet.*
26 Primula veris. *Common Cowflips or Paigles.*
27 Rumex fcutatus. *Round-leaved Sorrel.*
28 Rumex acetofa. *Common Sorrel.*
29 Salicornia europæa. *Jointed Glafswort or Saltwort.*
30 Scandix cerefolium. *Common Chervil.*
31 Scandix odorata. *Sweet Cicely.*
32 Sedum reflexum. *Yellow Stone Crop.*
33 Sedum rupeftre. *St. Vincent's Rock Stone Crop.*
34 Sifymbrium nafturtium. *Water-crefs.*
35 Sinapis alba. *White Muftard.*
36 Tanacetum balfamita. *Coftmary.*
37 Valeriana locufta. *Lambs Lettuce.*
38 Veronica beccabunga. *Brooklime.*
39 Ulva lactuca. *Green Laver.*

1 The ufe of the leaves of *Parfley* is well known

known in the kitchen, and the virtues of
the plant have been mentioned before; but
it may not be amiſs to obſerve farther, that
ſome farmers cultivate whole fields of this
plant, for the uſe of their ſheep, it being a
ſovereign remedy to prevent them from the
rot, provided they are fed with it twice or
thrice a week.　But this cannot be practiſed
where hares and rabbits abound, for theſe
creatures are ſo fond of it that they will
make long excurſions to get at it; and in a
ſhort time will deſtroy a large crop.

2 The *Allium cepa* too has been mentioned
in a former Chapter, and ſtands here only
on account of its leaves being in common
uſe among other cold ſallad herbs.

3 ALLIUM ſchœnopraſum. *Cives. Lin.
Sp. pl.* 432.
Porrum ſectivum juncifolium. *Bauh.
Pin.* 72.
This is an inhabitant of Siberia, and is a
very ſmall plant compared with the former,
the leaves and ſtems ſeldom exceeding ſix
inches in length, and the roots never producing any bulbs.　The leaves are awl-
ſhaped, hollow, and the ſtem naked.　It
was formerly in great requeſt for mixing
with ſallads in the ſpring, but has been little
regarded lately.　Its taſte, ſmell, and virtues
are

are much the fame as thofe of the common
Onion.

4 ALLIUM oleraceum. *Wild Garlick.*
Lin. Sp. pl. 429.
Allium montanum bicorne, flore exalbido.
Bauh. Pin. 75.

This grows in the paftures and corn-fields
in Effex, and fome other parts of England.
It hath a fmall, white, bulbous root, which
fends up a ftraight, round ftalk, about half
a yard high, furnifhed with a few rough,
pale green leaves, round on one fide, and
deeply furrowed on the other. The ftem
iffues from a horned fpatha or fheath, and is
terminated by an umbel of whitifh green
flowers, ftriped with purple.

The roots and leaves are ufed in Sweden
the fame as thofe of the common *Onion* are
here.

5 ARTEMISIA dracunculus. *Taragon.*
Lin. Sp. pl. 1189.
Dracunculus hortenfis. *Bauh. Pin.* 98.

This is a native of Siberia and other northern parts of Europe. It hath a woody
fort of root, compofed of a multitude of
fibres, and fends up feveral round, crooked,
branched ftalks, about two feet high, irregularly fet with long, narrow, fmooth,
lance - fhaped leaves, without footftalks;
thefe have a tafte and fmell almoft peculiar

to

to themfelves, but which are exceedingly grateful to many. The flowers are produced in clofe, flender panicles at the tops of the branches, and are of an herbaceous colour.

The leaves of this plant make an excellent pickle, which in the opinion of many people is not to be equalled by any other.

6 ALSINE media. *Chickweed. Lin. Sp. pl.* 389.

This is a fmall annual plant, and a very troublefome weed in gardens. The ftalks are weak, green, hairy, fucculent, branched, about eight inches long, and lodge on the ground. The leaves are numerous, nearly oval, fharp-pointed, juicy, of the colour of the ftalks, and ftand on longifh footftalks, having membranous bafes, which are furnifhed with long hairs at their edges. The flowers are produced at the bofoms of the leaves on long, flender peduncles; they are fmall and white, confift of five fplit petals each, and contain five ftamina and three ftyles.

The leaves of this plant have much the flavour of Corn-Sallad, and are eaten in the fame manner. They are deemed refrigerating and nutritive, and an excellent food for thofe of a confumptive habit of body. The plant formerly ftood recommended in the fhops as a vulnerary.

7 BORAGO

7 BORAGO officinalis. *Borage.* *Lin.*
Sp. pl. 197.

Bugloffum latifolium, Borago. *Bauh.*
Pin. 259.

This is an annual, and grows plentifully
by road fides, and other uncultivated places.
It alfo is cultivated in gardens, in order to
have it at hand to mix with ftuffing herbs,
and to put into cool tankards, whereby the
plant is fufficiently known. The whole is
fuppofed cordial and exhilarating, but for
what reafon is difficult to guefs, as neither
the fmell or tafte countenance any fuch
properties.

8 CACALIA ficoides. *Fig Marigold-*
leaved Cacalia. *Lin. Sp. pl.* 1168.

This is a fhrubby plant, and a native of
Æthiopia. From the root rife feveral round
ftalks, to the height of feven or eight feet;
thefe are hard and woody below, but tender
and fucculent upward, where they fend out
many irregular branches, which are furnifhed with lance-fhaped, compreffed, flefhy
leaves, ending in acute points, covered with
a whitifh farina or meal, that eafily comes
off when touched. The flowers are produced at the extremity of the branches, in
fmall umbels; they are compofed of many
white, tubular, hermaphrodite florets, ftanding in a common cylindrical calyx, are cut
at their brims into five parts, and each contains

tains five fhort flender ftamina, and one ftyle, faftened to an oblong germen, which becomes an oblong feed, crowned with long down.

The leaves of this plant are pickled by the French, who efteem them much; and in doing this they have a method of pre-ferving the white farina upon them, which greatly adds to the beauty of the pickle when brought to table.

9 CICHORIUM endivia. *Endive. Lin. Sp. pl.* 1142.

Cichorium latifolium five Endivia vulgaris. *Bauh. Pin.* 125.

The *Endive* and its varieties have been fo long cultivated in England, and other parts of Europe, that it is impoffible to tell with certainty what country claims it as a native. The plant is well known in the gardens, and its ufes in the kitchen.

In regard to its phyfical properties it is counted detergent, aperient, and attenuating, tending rather to cool than heat the body. By opening obftructions of the liver, it gives relief in the jaundice; and by its de-terging quality, it is ferviceable in fcorbu-tic habits.

10 COCHLEARIA officinalis. *Scurvygrafs. Lin. Sp. pl.* 903.

Cochlearia folio fubrotundo. *Bauh. Pin.* 110.   **This**

This is found wild in the Marſhes near the northern coaſts of England, but it is probable it was at firſt introduced into our gardens from Holland, where it grows very plentifully. It is an annual plant, with a ſmall fibrous root, from which come many roundiſh, fleſhy, ſhining green leaves, a little waved on their edges, and are ſupported on long foot-ſtalks. Among theſe riſe ſeveral pale green, round ſtalks, a little branched towards their tops, and having a few oblong, ſharp - pointed, light-green leaves, ſtanding on them by pairs. The ſtalks riſe to about a foot high, producing various bunches of flowers, conſiſting of four ſmall white petals each, placed nearly at right angles with each other, and ſurrounding ſix ſtamina, four of which are longer than the reſt. The germen is nearly heart-ſhaped, and becomes a roundiſh ſeed-veſſel, having two cells, ſeparated by a thin membrane, in each of which are contained four or five round ſeeds.

The leaves of this plant are exceedingly pungent, therefore the beſt way of eating them is between bread and butter, as by this means they are rendered leſs offenſive to the palate, and their whole virtues, which are very conſiderable, are taken into the ſtomach. Uſed any way they divide viſcid juices, open obſtructions, ſcour the glands, and become a ſovereign remedy againſt the

H   ſcurvy;

scurvy; all which have juftly obtained the plant the name of Scurvy grafs. There is a conferve, and a plain fpirit of it kept in the fhops, both which are in great efteem, but they are far inferior, as anti-fcorbuticks, to the frefh leaves, eaten as above directed; frequently ufed in this manner they muft prove beneficial in all cold phlegmatic conftitutions, and cleanfe the fkin of fcabs, and other cutaneous erup-tions.

11 Erysimum alliaria. *Jack by the Hedge. Lin. Sp. pl.* 922.

This is a very common plant among bufhes and in hedge-rows. It is a peren-nial, and hath a long, whitifh root, divid-ed into feveral parts. The radical leaves rife in a clufter, upon long, flender foot-ftalks; they are heart-fhaped, of a light yellowifh green colour, about three inches broad, and crenated on their edges. The ftem is erect, firm round, fometimes a little branched, about a yard high, and furnifhed with leaves like thofe below, but fmaller. It terminates in a racemus of whitifh flowers, having four petals each, including fix fta-mina, two of which are fhorter than the reft, and one very fhort ftyle. The fuc-ceeding pods are long, flender, all the way of a thicknefs, and contain many fmall black-ifh iſh feeds. The whole plant has the ſmell and taſte of Garlick.

The poor people in the country eat the leaves of this plant with their bread, and on account of the reliſh they give, call them *Sauce-alone*. They alſo mix them with Lettuce, uſe them as a ſtuffing herb to pork, and eat them with ſalt-fiſh. The plant was in high eſteem formerly as an attenuater, and powerful expectorant, and held immediately uſeful in aſthmas, and diſtillations of rheum upon the lungs.

12 ERYSIMUM barbarea. *Winter-creſs. Lin. Sp. pl. 922.*

Eruca lutea latifolia ſive barbarea. *Bauh. Pin. 98.*

The *Winter-creſs* grows plentifully on moiſt banks and by ditches. It is a perennial, and hath a long thickiſh root, furniſhed with a few fibres. The bottom leaves are cut into four or five pair of lobes, like pinnæ, with a large roundiſh one at the end. Among theſe come ſeveral flower-ſtems, about half a yard high, irregularly ſet with leaves like thoſe from the root, but they are ſmaller. The ſtems divide into many branches, terminated by looſe ſpikes of ſmall yellow flowers, having four petals each, which include ſix ſtamina, two ſhorter than the reſt, and one ſtyle. The ſucceeding pods are long and ſlender. There is a

H 2   beautiful

beautiful variety of this plant in gardens, with a double flower, and is generally called the yellow Rocket.

The leaves were formerly mixed with fallad herbs, but their having rather a rank fmell, and no very agreeable flavour, are now neglected here, though in Sweden they ftill retain a place at table.

The plant is a powerful ' antifcorbutic, and no way inferior to the Water-crefs.

13 Fucus faccharinus. *Sea Belts. Lin. Sp. pl.* 1630.

Fucus alatus five phafnagoides. *Bauh. Pin.* 364.

This is a weed that grows upon rocks and ftones by the fea-fhore. It confifts of a long, fingle leaf, having a fhort roundifh foot-ftalk, the leaf reprefenting a belt or girdle.

14 Fucus palmatus. *Handed Fucus. Lin. Sp. pl.* 1630.

This grows alfo in the fea, and confifts of a thin, lobed leaf, in the form of a hand.

15 Fucus digitatus. *Fingered Fucus. Hud. Flo. Ang.* 579.

Fucus arboreus polyfchides edulis. *Bauh. Pin.* 364.

This grows likewife upon ftones and rocks in the fea near the fhore. It hath feveral

plain,

plain, long leaves or finufes, fpringing from a round ftalk, in the manner of fingers when extended.

16 Fucus efculentus. *Edible Fucus. Hud. Flo. Ang. 578.*

Mr. Hudfon has made this a diftinct fpecies, but Linnæus included it under his *faccharinus.* It grows plentifully in the fea, near the fhores of Scotland, and alfo thofe of Cumberland. This hath a broad, plain, fimple, fword-fhaped leaf, fpringing from a pinnated ftalk. All thefe four fpecies are collected by the failors, and people along the fea-coafts, as fallad herbs, and are efteemed excellent antifcorbuticks. The leaves of the *faccharatus* are very fweet, and when wafhed and hanged up to dry, will exude a fubftance like that of fugar.

17 Hypochæris maculata. *Spotted Hawkweed. Lin. Sp. pl.* 1140.

Hieracium alpinum latifolium hirfutia incanum, flore magno. *Bauh. Pin.* 128.

This is a perennial plant, and a native of England. The root is compofed of a multitude of fibres, from which fpring a clufter of large, oval, hairy, deep green, fpotted leaves, having fharp teeth, fet at confiderable diftances along their margins. The ftalk rifes in the midft of thefe, with a bunch of feffile leaves near its bafe; it is up-

H 3 right,

right, firm, and naked from thence to the top, where moftly ftands only one large, gold-coloured compound flower, having an imbricated calyx, and confifting of herma-phrodite, tongue-fhaped florets, cut into five teeth at their brims, and each contain-ing five fhort, hairy ftamina and one ftyle.

The leaves are eaten as thofe of *Lettuce*, and are deemed cooling; they are alfo boil-ed in broths.

18 LACTUCA fativa.   *Garden Lettuce.* *Lin. Sp. pl.* 1118.

This hath been fo long cultivated in gar-dens, that its native place of growth is not known. The varieties of it are very nume-rous; Dr. Boerhaave has given a lift of 47 that were growing in the Botanic Garden, at Leyden, in the year 1720, and we have near a fcore at this time cultivated in Eng-land. *Lettuce* is a cooling, emolient, laxa-tive plant, but like moft lactefcent ones has a narcotic quality, as any one may per-ceive who eats plentifully of it.

19 LEONTODON taraxacum.   *Dandelion.* *Lin. Sp. pl.* 1122.

Dens leonis, latiore folio.   *Bauh. Pin.* 126.

This is a moft troublefome weed to far-mers and gardeners, for when it is once fixed in their grounds, it is no eafy matter

to

to eradicate it, by reafon its downy feeds fly to all parts and vegetate on any foil; hence the plant is fo well known as to render a defcription of it ufelefs.

The young tender leaves are eaten in the fpring as Lettuce, they being much of the fame nature, except that they are rather more detergent and diuretic. Boerhaave greatly recommended the ufe of *Dandelion* in moft chronical diftempers, and held it capable of refolving all kinds of coagulations, and the moft obftinate obftrudions of the vifcera, if it were duly continued. For thefe purpofes the ftalks may be blanched and eaten as Celery.

20 LEPIDIUM fativum. *Garden Crefs. Lin. Sp. pl. 899.*

Nafturtium hortenfe vulgatum. *Bauh. Pin. 103.*

This is an annual plant, and a native of Germany. The leaves are long, narrow, and deeply cut into irregular fegments. The ftalk is round, firm, upright, about two feet high, of a whitifh green colour, a little branched towards the top, and is all the way furnifhed with many jagged leaves.

The flowers come out in bunches at the tops of the branches, each confifting of four fmall, white petals, including fix ftamina, four longer than the reft, and one ftyle; thefe

thefe are fucceeded by a kind of heart-fhaped pods, containing brown feeds.

The plant is now generally fown in gardens for a fpring fallad, and perhaps a better can fcarcely be cultivated. It is of a warm, ftimulating nature, having much the fame qualities as the Watercrefs, but is lefs pungent. There is a variety of this with curled leaves, which has the fame properties with the original, but is more ufed for garnifhing difhes than fallading.

21 LEPIDIUM Virginicum. *Virginian Sciatic Crefs. Lin. Sp. pl.* 900.

Though the *Virginicum*, as its name expreffes, grows in Virginia, yet it is alfo an inhabitant of feveral of the Weft-India Iflands, and efpecially of Jamaica.

It is an annual, and fends forth a very branched ftalk, about half a yard high, fet with narrow, winged leaves, the lobes of which are finely ferrated.

The flowers come out in the manner of thofe of the *fativum,* but fome of them have only three ftamina.

The people in America gather the plants, and eat the leaves as we do thofe of the Garden Crefs.

22 MENTHA fativa. *Marfh, or Curled Mint. Lin. Sp. pl.* 805.

Mentha crifpa vefticillata. *Bauh. Pin,* 227.

The

The *Mentha sativa* grows wild by marshes and rivulets. It is a perennial, and creeps much by the roots, as most of the Mints do. The stalks are about half a yard high, square, of a purplish colour, throw out many shoots from the bosoms of the leaves, and are generally bent near their base.

The leaves are oval, serrated, wrinkled, of a pale green, and often curled at their edges.

The flowers are purple, and come out in whorles at the joints of the branches. The whole plant has a very pleasant smell.

23 MENTHA viridis. *Spear Mint. Lin. Sp. pl.* 804.

Mentha angustifolia spicata. *Bauh. Pin.* 227.

The *viridis* too grows naturally by runs of water. This is a taller plant than the former, having a firm, square, upright stalk, two feet or more high, sending out many branches from the bosoms of the leaves.

The leaves are of a lively green colour, long, narrow, sharp pointed, and deeply serrated at the edges.

The flowers stand at the tops of the stalks, in slender spikes, and are of a bright red colour.

Though this is the sort most cultivated for culinary uses, yet to many palates it is far inferior in pleasantness to the former.
They

They are much alike in their virtues, being
ſtomachic and carminative.

24 OXALIS acetoſella. *Wood Sorrel.*
Lin. Sp. pl. 620.

Trifolium ácetoſum vulgare. *Bauh. Pin.*
330.

The *Oxalis acetoſella* is a neat little plant,
common in our woods. It hath a ſlender,
creeping, irregular root, hung with many
fibres. The leaves riſe in little cluſters;
they are heart-ſhaped, and are joined by
their points three together at the top of a
long, weak, reddiſh foot-ſtalk, with their
broad ends hanging downward. Their colour is a yellowiſh green, and they are a little hairy.

Among theſe, and immediately from the
root, come the flower-ſtalks, each ſupporting a pale fleſh-colour, bell-ſhaped flower,
ſnipped into five ſegments almoſt down to
the baſe, and containing ten hairy, erect ſtamina, and five ſlender ſtyles.

The leaves of this plant afford one of the
moſt grateful acids of any in nature, far
preferable to that of the common garden
Sorrel, and therefore is more eligible for
mixing with ſallads. They are cooling,
and ſerviceable againſt inflammatory diſorders. Beaten with ſugar they make an ele-
gant conſerve; and boiled with milk form a
moſt

moſt agreeable whey, which is good for opening obſtructions of the viſcera.

25 POTERIUM ſanguiſorba. *Burnet.*
Lin. Sp. pl. 1411.
Pimpinella Sanguiſorba minor hirſuta.
Bauh. Pin. 160.

The *Poterium ſanguiſorba* is common in chalky grounds, and hilly paſtures. It is ſo frequently cultivated in gardens, that to deſcribe it would be unneceſſary; its uſes in the kitchen too are generally known. It is counted cordial and ſudorific, and on that account is often put into cool tankards. It evidently has an aſtringent quality, and thereby is ſerviceable againſt dyſenteries.

26 PRIMULA veris. *Cowſlips. Lin. Sp. pl.* 204.
Verbaſculum pratenſe odoratum. *Bauh. Pin.* 241.

Linnæus makes the Common *Cowſlip*, the great *Oxlip*, and the Common *Primroſe*, only variations of one and the ſame ſpecies, but in this he is certainly wrong, as the *Primroſe* is evidently a diſtinct one. They are all too well known to require any deſcriptions, and their leaves may be uſed promiſcuouſly. As to their being eſculent, they are only ſo as they enter into compoſition with other herbs, in the ſtuffing of meat. From the flowers, indeed, of the
Cowſlip

Cowflip a very good wine is made, but it is not equal to that drawn from *Clary*.

27 RUMEX fcutatus. *Round-leaved Sor-rel. Lin. Sp. pl.* 480.

Acetofa rotundifolia hortenfis. *Bauh. Pin.* 114.

The *Rumex fcutatus* is a native of Swit-zerland. It is a perennial, and hath a creeping, fibrous root, which fends forth many leaves on long foot-ftalks; thefe are hollow in the middle like a fpoon, and are betwixt the fhape of an heart, and that of the head of an arrow.

The ftalk rifes a foot or more high, fet with leaves till near the top, where it breaks into flender fpikes of brownifh green flow-ers, containing fix ftamina and one ftyle each.

The leaves having a very pleafant, acid tafte, the plant is frequently raifed in our gardens to mix with fallad herbs.

28 RUMEX acetofa. *Common Sorrel. Lin. Sp. pl.* 481.

Acetofa pratenfis. *Bauh. Pin.* 14.

The *Acetofa* grows very common in our woods and meadows. This too is a peren-nial, and from a long, yellowifh, woody root, fends up a curved, channelled, reddifh ftalk, about two feet high, confifting of a few joints, with a long, arrow-fhaped leaf

5   at

at each. The leaves at the bottom of the
ftalk have long foot-ftalks, but thofe towards the top ftand clofe, and embrace the
ftalk with their bafe. At the top of the
ftalk comes forth a branched panicle of fmall
reddifh flowers, refembling thofe of Dock.
There are feveral wild varieties of this
plant.

The leaves have little or no fmell, but
when chewed have a reftringent acid tafte.
Their medicinal effects are to cool, quench
thirft, and promote the urinary difcharge.
They are frequently mixed with fallad herbs
the fame as the former.

The Irifh, who are particularly fond of
acids, eat the leaves with their milk and
fifh; and the Laplanders ufe the juice
of them as rennet to their milk. The
Greenlanders cure themfelves of the Scurvy
with the juice of Scurvy grafs and this
mixed; and Dr. Boerhaave recommends a
decoction of the leaves as an efficacious remedy againft inflammatory diforders.

29 SALICORNIA europæa, vel herbacea.
Jointed Glaſſwort. Lin. Sp. pl. 5.
This is an annual plant, and grows plentifully in the falt marſhes, in many parts of
England. It varies very much in the nature of its growth, infomuch that different
writers on Botany have made three or four
different fpecies of it. It hath fucculent,
jointed,

jointed, branched ftalks, which in fome plants, trail upon the ground, and in others ftand upright. The flowers are produced at the ends of the joints, towards the extremity of the branches; thefe are fo fmall as fcarce to be difcerned with the naked eye.

This plant is gathered by the country people, and fold about for the true *Samphire*, but it is very different from that plant. (See *Crithmum maritimum*). This, however, makes an excellent good pickle, which renders the cheat the lefs to be regretted. They alfo cut the plants up towards the latter end of fummer, when they are full grown, and after having dried them in the fun, they burn them for their afhes, which are ufed in making of glafs and foap. The *Sal Kali* of the fhops was formerly drawn from the afhes of this plant only, but now from fundry forts of herbs. The manner of obtaining the *alkali*, is to dig a hole, and lay laths acrofs it; on thefe they pile the herbs, and having made a fire under the laths, the herbs are fuffered to burn till their liquor drops from them to the bottom of the hole, where it hardens, and turns of a blackifh afh colour, retains a faltifh tafte, and is very fharp and corrofive.

30 SCANDIX cerefolium. *Common Chervil. Lin. Sp. pl.* 368.

2

Chæro-

Chærophyllum fativum. *Bauh. Pin.* 152.

The *Scandix cerefolium* is a fmall annual plant, with winged leaves, fomewhat refembling Parfley at firft, but of a yellower colour, and generally turning reddifh as they grow old.

The ftalks are upright, hollow, ftriated, much branched, fwelled in knobs under their joints, and have leaves on them like thofe from the root, except being divided into narrower fegments.

The flowers come out in umbels at the tops of the branches; they are fmall and white, and are fucceeded by longifh-oval, fhining, fharp-pointed feeds, of a dark brown colour. It is a native of the Auftrian Netherlands.

The plant is grateful to the palate, and is much cultivated by the French and Dutch, who are fo very fond of it, that they have hardly a foup or fallad but the leaves of Chervil make part of it. The ancients had the plant in the higheft efteem, and held it capable of eradicating moft chronical diftempers; it being mild, aperient and diuretic, working without irritation, yet breaking fabulous concretions, and allaying heat in the urinary paffages, whereby it proved particularly ferviceable in dropfies and the gravel. Some of them have gone fo far as to affert, that if thefe diforders would not yield to a conftant ufe of

this

this plant, they were fcarce curable by any other medicine.

31 SCANDIX odorata. *Sweet Cicely. Lin. Sp. pl.* 368.

Myrrhis major, Cicutaria odorata. *Bauh. Pin.* 160.

The *odorata* is a perennial, and a very large plant compared to the former. It has a thick white root, compofed of many fibres, which have a fweet, aromatic tafte. This fends forth feveral large, winged leaves, bearing fome refemblance to Fern, but they have often white fpots upon them.

The ftalk rifes four or five feet high, is hairy, fiftulous, and terminated by large umbels of white flowers, having five irregular petals each. Thefe are fucceeded by long, angular, deep-furrowed feeds, which when chewed, have a fweet, aromatic flavour like Anife-Seeds.

The leaves have nearly the fame flavour, and are employed in the kitchen as thofe of the *cerefolium.* The green feeds chopped fmall and mixed with Lettuce or other cold fallads, give them an agreeable tafte, and render them warm and comfortable to the ftomach. The plant is a native of Italy.

32 SEDUM reflexum. *Yellow Stone-Crop. Lin. Sp. pl.* 618.

Sedum

Sedum minus luteum, folio acuto. *Bauh.
Pin.* 283.

The *Sedum reflexum* is common upon old
walls and rocks, where it creeps much by
the roots, fending forth many weak, flender
fhoots, fet all round with fucculent, half-
round, fharp-pointed leaves. The flower-
ftalks rife from the fides of thefe fhoots to
about nine inches high, and are furnifhed
with leaves like the former, the bafes of
which turn a little upwards, and are moftly
tinged with red.

The ftalks are terminated by an umbel of
yellow flowers, confifting of five fharp-
pointed petals, which ftand horizontally in
form of a ftar, and contain ten awl-fhaped
ftamina, with five flender-ftyles each. Before the flowers come out, the rays of the
umbel are rolled up in manner of the Ionic
volute. There is however a variety *(Sedum
minus hæmatoides)* with ftraight rays.

The plant is cultivated by the Dutch,
who mix the leaves amongft their fallads.
They have a fubaftringent tafte.

33 SEDUM rupeftre. *St. Vincent's Rock
Stone-crop. Lin. Sp. pl.* 618.

The *rupeftre* grows upon St. Vincent's
rock, near Briftol. The firft fhoots are
branched, thickly covered with oblong,
flefhy leaves, and lodge upon the ground.
Among thefe rife the ftems to five or fix

I   inches

inches high, fet with awl-fhaped leaves, each having a fhort, loofe membrane at its bafe, which falls off upon being touched. They are of a fea-green colour, and rather rigid.

The flowers terminate the ftalks in round-ifh bunches, and are of the form, and nearly of the colour of the *reflexum*.

This plant too is cultivated by the Dutch, who ufe the leaves and tender tops as they do thofe of the former.

34 SISYMBRIUM nafturtium.   *Water-crefs.   Lin. Sp. pl.* 916.

Nafturtium aquaticum fupinum.   *Bauh. Pin.* 104.

The *Sifymbrium nafturtium* is common in our rivulets and water-ditches, and is fo well known and fo much in ufe, that many families in the country have it conftantly at their tables two or three months in the year. It is a good diuretic, a powerful re-folver of phlegmatic juices, and thereby a fovereign remedy againft the fcurvy.

35 SINAPIS alba.   *White Muftard.   Lin. Sp. pl.* 933.

Sinapi apii folio.   *Bauh. Pin.* 99.

This grows fpontaneoufly on hedges and the borders of fields. It fends up a branched ftalk about two feet high, furnifhed with rough leaves, deeply jagged down to the midrib.

midrib. The branches are terminated by loofe fpikes of fmall yellow flowers, each having four petals placed in form of a crofs. Thefe are fucceeded by hairy, rough pods, with long, flat beaks. The plant is now much cultivated in gardens, for a fallad-herb in the fpring.

In regard to its medicinal properties, it is nearly of the nature of the Watercrefs, and ftands recommended as good for exciting the appetite, promoting digeftion, atte-nuating vifcid juices, and thereby promoting the fluid fecretions.

36 TANACETUM balfamita.   *Coftmary.* *Lin. Sp. pl.* 1184.

Mentha Hortenfis corymbifera.   *Bauh.* *Pin.* 226.

The *Tanacetum balfamita,* is a perennial plant, and a native of the fouthern parts of France and Italy. It hath a creeping fibrous root, which produces many oval, greyifh-green leaves, finely ferrated at the edges, and ftanding upon long footftalks.

Among thefe rife feveral round, green, branched ftems, to above half a yard high, with fuch leaves thereon as thofe from the root, but fmaller. The branches are ter-minated by bunches of yellow flowers re-fembling thofe of Tanfey.

The whole plant has an agreeable fmell, which to many is far preferable to any of

the

the Mints. It was formerly cultivated in gardens for the purpofe of mixing with fallads, and it is a pity it is not continued, as from its fenfible qualities it feems fuperior to many aromatic plants now in credit.

37 Valeriana locufta. *Lamb's Lettuce.* *Lin. Sp. pl.* 47.

The *Valeriana locufta* is found wild in fields, on banks, and old walls. It is generally known by being cultivated in gardens under the name, Corn-fallad. The leaves ought to be cut young for fallading, otherwife they have a difagreeable bitter tafte. It is a plant that varies much by foil and fituation. Linnæus has fix varieties of it, yet he has not enumerated them all.

38 Veronica beccabunga. *Brooklime.* *Lin. Sp. pl.* 16.

The *Veronica beccabunga* is frequent in fhallow waters, and by the fides of brooks. It hath a long creeping root, which fends clufters of fibres into the mud. From this come feveral weak fhoots, that ftrike root frequently as they trail along. Thefe are round, of a pale green colour, and fpungy fubftance, as are the ftalks, and fet at their joints with thick, fmooth, oval leaves, about an inch long, ftanding oppofite each other, clofe to the ftalks.

The flowers come out in long, flender

5

bunches only at the bofoms of the leaves, for the main ftems are always terminated by fmall clufters of leaves, not flowers. Each flower is compofed of one fine blue petal, which fpreads flat, and is cut at the brim into four unequal fegments. In the centre are two ftamina and one ftyle, and it is fucceeded by a fmall heart-fhaped pod, having two cells.

The leaves are very pungent and bitterifh, yet are eaten by many with bread and butter. The plant is in the higheft efteem as an antifcorbutic, and is faid even to furpafs the Watercrefs; this may not be conceit only, by reafon it has the pungency of the latter, and is much more aftringent. The juice ftands in the firft clafs of the fweeteners of the blood. The country people cure green wounds with no other application than thefe leaves frefh gathered.

39 ULVA lactuca. *Green Laver.* *Lin. Sp. pl.* 1632.

Mufcus marinus lactucæ fimilis. *Bauh. Pin.* 364.

The *Ulvalactuca* is a broad, membranaceous leaf, or rather a collection of fuch leaves, growing from each other. It is found on rocks and ftones in the fea, and often upon oyfter-fhells, and has fome refemblance to curled Lettuce, whence the name *lactuca*. The failors and inhabitants along the coafts

I 3   devour

devour it with great avidity, efteeming it good againft the fcurvy. It is pleafant to the palate, and gently laxative.

S E C T.   II.

Boiling Sallads.

1 AMARANTHUS oleraceus. *Efculent Amaranth.*

2 Arum efculentum. *Indian Kale.*

3 Atriplex hortenfis. *Garden Orach.*
———— *hortenfis nigricans.* Dark green Garden Orach.
———— *hortenfis rubra.* Red Garden Orach.

4 Anethum fœniculum. *Common Fennel.*
———— *dulce.* Sweet Fennel.

5 Braffica oleracea. *&c. Cabbages.*

6 Braffica napus. *Navew or Colewort.*

7 Chenopodium bonus Henricus. See Chap. II.

8 Cnicus oleraceus. *Round-leaved Meadow Thiftle.*

9 Corchorus olitorius. *Common Jews Mallow..*

10 Crambe maritima. *Sea Colewort.*

11 Jatropha maniot. *Caffava.*

12 Malva rotundifolia. *Dwarf Mallow.*

13 Mentha viridis. *Spear Mint.* See Sect. I.

14 Phytolacca

14 Phytolacca decandra. *American Night-*
    ſhade.
15 Ranunculus ficaria. *Pilewort.*
16 Raphanus ſativus. *Common Radiſh.*
17 Salvia ſclarea. *Garden Clary.*
18 Spinacia oleracea. *Common Spinach.*
    —— *oleracea glabra.* Smooth Spinach.
19 Thea bohea, *Bohea Tea.*
20 Thea viridis. *Green Tea.*
21 Urtica dioica. *Common Stinging Nettle.*

    1 AMARANTHUS oleraceus. *Eſculent*
Amaranth. Lin. Sp. pl. 1403.
    Blitum album majus? *Bauh. Pin.* 118.
    This is a native of India, and an annual.
It ſends forth many large, rough, oval,
brittle leaves, reſembling thoſe of the White
Beet, but more obtuſe, and ſnipped at their
apex. Among theſe riſes the ſtalk to much
the ſame height as that of the particoloured
Amaranthus, and is terminated by a pale,
glomerated ſpike, which is longer than
thoſe that terminate the branches. Some
few of the flowers have five ſtamina, but the
much greater part have only three,
    The leaves of this are boiled in India the
ſame as Cabbage is here. Though Linnæus
by his trivial name has pointed this ſpecies
out in particular for an eſculent one, yet the
leaves of ſeveral others of the genus are alſo
eaten.

2 ARUM efculentum. *Indian Kale.*

This having been defcribed in the firft
divifion, it remains only to obferve here,
that the Indians boil the leaves as a fallad,
and efteem them very wholefome.

3 ATRIPLEX hortenfis. *Garden Orach.*
Lin. Sp. pl. 1493.

This is an annual, and a native of Tar-
tary. It hath almoft triangular, obtufe
pointed leaves, ftanding oppofite, on long,
flender footftalks. Thefe are generally covered at their bafe with a mealy duft, as is
the upper part of the ftalk alfo. It was
much cultivated in the Englifh gardens formerly, but now its place is chiefly fupplied
by *Spinach.* The French, however, ftill
efteem it, and there are fome palates among
us that prefer it to *Spinach.* It is of a cooling, laxative nature, and an excellent fallad
for thofe of a coftive habit of body. The
names of its varieties are fufficient defcriptions of them.

4 ANETHUM fœniculum. *Fennel. Lin.*
Sp. pl. 377.

Fœniculum dulce. *Bauh. Pin.* 147.

This is frequently found wild in many
places; neverthelefs it certainly is not a
native here, but was originally brought hi-
ther from Spain or Germany. The ufe of
its leaves is too well known in the kitchen
to have any thing faid about it. In regard

to

to the virtues of the plant, it is of a warm active nature, and good to expel flatulencies. The variety, called sweet *Fennel*, differs much from the common, its leaves being larger, and slenderer, its stalks shorter, the seeds longer, narrower, of a lighter colour, sweet, and mostly bent inwards.

This last is greatly cultivated in Italy and Germany, whence the seeds are imported.

5 Brassica oleracea, &c. *Cabbages.*

Cabbages of all kinds are supposed to be hard of digestion, to afford but little nourishment, and to produce flatulencies; but they seem to have this effect only on weak stomachs, for there are many who will feed heartily upon them, and feel none of these inconveniencies. Few plants run into a state of putrefaction sooner than these, and therefore they ought to be used when fresh cut. In Holland and Germany they have a method of preserving them, by cutting them in pieces, and sprinkling salt and some aromatic herbs among them; this mass is put into a tub, where it is pressed close, and left to ferment, and then it is called *Sour Crout*. Thus managed it is sent on ship-board in barrels, and proves a refreshing dish to the sailors; or at least, it is certainly the means of keeping them from the scurvy.

6 Brassica

6 BRASSICA napus. *Colewort. Lin. Sp. pl.* 931.

Napus fylveftris. *Bauh. Pin.* 95.

This is a biennial plant, and is frequently found wild in corn-fields. It hath a long white root, which fends forth feveral pale green jagged leaves. Among thefe rifes the ftalk, to three or four feet high, irregularly fet with lance-fhaped leaves, flightly notched at their edges, having broad bafes embracing the ftem. The flowers are yellow, ftand in tufts at the extremities of the branches, confift of four petals each, and are fucceeded by long pods.

There are many varieties of this plant cultivated in gardens for winter and fpring fallads, and are called Collets or Coleworts*. In fome counties whole fields are fown with Navew as feed for cattle, or for the feed; for it is from thefe feeds that the Rape oil is drawn. All domeftic fowls, and feveral wild ones, efpecially pheafants and partridges, are very fond of thefe feeds, and will deftroy a great part of a crop, unlefs it be well guarded.

8 CNICUS oleraceus. *Round-leaved Meadow Thiftle. Lin. Sp. pl.* 1156.

* Thefe forts of Coleworts are now almoft banifhed by the gardeners, and inftead thereof they fow the feeds of the *Yorkfhire* or *Sugar-loaf Cabbage*, calling the young plants thus raifed, Coleworts, though very improperly.

Carduus

Carduus pratenfis latifolius. *Bauh. Pin.*
376.

This plant is a native of the northern
parts of Europe, where the inhabitants boil
the leaves as we do Cabbage. It is a perennial, and fends forth large oblong leaves,
deeply cut at their edges into various fegments, which are ferrated, and furnifhed
with whitifh green, tender fpines. The
ftalk rifes three or four feet high, breaking
into branches, which are fet with leaves, at
whofe bofoms the flowers are produced on
long peduncles. Thefe are compofed of all
hermaphrodite florets, furrounded by green,
prickly fcales, which are nipped up. The
feeds ftand fingly upon a flat, hairy receptacle, and are crowned with a feathery
down.

9 CORCHORUS olitorius. *Common Jews
Mallow. Lin. Sp. pl.* 746.
Corchorus Plinii. *Bauh. Pin.* 317.

This is an annual, and a native of Afia,
Africa, and America. It rifes with a round,
ftriated, upright, branched ftalk, to near
two feet, which is furnifhed with leaves
differing in fhape; fome being oval, fome
cut off ftraight at their bafe, and others almoft heart-fhaped. They are of a deep
green colour, and have a few teeth on the
margins of their bafe, that end in briftly,
reflexed, purplifh filaments. The flowers
come

come out at the fides of the branches, oppofite to the leaves; they ftand fingly on
very fhort peduncles, are compofed of five
fmall yellow petals, and a great number of
ftamina, furrounding an oblong germen,
which becomes a long, rough, fharp-pointed
capfule, opening in four parts, each filled
with greenifh, angular feeds.

This plant is fown by the Jews about
Aleppo, and is therefore called *Jews Mallow.* The leaves are a favourite fallad
among thefe people, and they boil and eat
them with their meat.

10 CRAMBE maritima. *Sea Colewort.*
Lin. Sp. pl. 937.
Braffica maritima monofpermos. *Bauh.
Pin.* 112.

This grows naturally on the fea coaft in
many parts of England. It hath a long,
thick, creeping root, divided into various
fibres, and fends up feveral fpacious, nearly
oval leaves, much jagged on their edges, of
a greyifh green colour, and flefhy fubftance.
In the centre of thefe rifes a round, whitifh,
upright ftalk, two feet or more high, dividing near the top into a few branches,
having a few feffile, oval leaves. The
branches are terminated by loofe bunches
of fmall white flowers, compofed of four
petals each in form of a crofs, and containing fix ftamina, two of which are fhorter
than

than the reſt, and one ſtyle. Theſe are ſucceeded by roundiſh capſules, about the ſize
of peas, each including one round ſeed.

The radical leaves being green all the
winter, are cut by the inhabitants where the
plants grow, and boiled as Cabbage, to
which they prefer them.

11 JATROPHA maniot. *Caſſava.*

The *Jatropha maniot* has been deſcribed in the firſt Chapter; its name is repeated
here, by reaſon the leaves are boiled and
eaten by the Indians, in the ſame manner
as *Spinach* is by us.

12 MALVA rotundifolia. *Dwarf Mallow.*
Lin. Sp. pl. 969.

Malva ſylveſtris, folio ſubrotundo. *Bauh.*
Pin. 314.

This is a ſmall ſort of *Mallow,* that grows
by old walls, and rude, uncultivated places.
From a long white root it ſends forth a
cluſter of pale green, roundiſh leaves, having
long footſtalks, and are coarſely crenated on
their edges. Among theſe iſſue many long,
ſlender, proſtrate ſtalks, plentifully furniſhed with ſuch-like leaves, ſtanding irregularly on them. The flowers come out at
the footſtalks of the leaves, and alſo at the
ends of the branches, on bending peduncles,
and each is compoſed of one pale fleſh-co-
loured petal, cut into five ſegments down

to

to the bafe, including many ftamina united below in form of a cylinder.

The leaves of this plant were formerly in great efteem as a fallad that would abate heat in the bowels, and obtund acrimonious humours; but at prefent it is totally neg-lected.

14 PHYTOLACCA decandra: *American Nightfhade. Lin. Sp. pl.* 631.

This, grows naturally in the province of Virginia, in America. It hath a thick, flefhy, perennial root; divided into feveral parts as large as middling Parfneps. From this rife many purplifh, herbaceous ftalks, about an inch thick, and fix or feven feet long, which break into many branches, irre-gularly fet with large, oval, fharp-pointed leaves, fupported on fhort footftalks. Thefe at firft are of a frefh green colour, but as they grow old they turn reddifh. At the joints, and divifions of the branches, come forth long bunches of fmall bluifh-coloured flowers, confifting of five concave petals each, furrounding ten ftamina and ten ftyles. Thefe are fucceeded by round depreffed ber-ries, having ten cells, each of which contains a fingle fmooth feed.

In Virginia and other parts of America the inhabitants boil the leaves, and eat them in the manner of Spinach. They are faid to have an anodyne quality, and the juice

of

of the root is violently cathartic. The
Portugueze had formerly a trick of mixing
the juice of the berries with their red wines,
in order to give them a deeper colour; but
as it was found to debafe the flavour, the
matter was reprefented to his Portugueze
Majefty, who ordered all the ftems to be cut
down yearly before they produced flowers,
thereby to prevent any further adulteration.

15 RANUNCULUS ficaria. *Pilewort*. *Lin*.
Sp. pl. 774.
Chelidonia rotundifolia minor. *Bauh*.
Pin. 309.
This is a perennial plant, and to be met
with on moift banks and in meadows. It
has a root compofed of many little tubercles
fufpended by fibres; which tubercles fomewhat refemble the outward piles, hence the
name of the plant. The leaves are triangular, heart-fhaped, of a fine glaffy green,
ftreaked in the middle with blackifh and
whitifh lines. The flower-ftems rife four
or five inches high, having many leaves at
their bafe, and each is terminated by one
yellow flower, confifting of feveral narrow,
fharp-pointed petals *, furrounding a great
many ftamina and ftyles. Thefe flowers
make no little part of the variegated covering
of meadows and moift paftures in the fpring.

* Thefe are fubject to vary, they being roundifh in fome
plants, and in fuch the leaves are moftly obtufe-angled.

There

There is a variety of this plant in gardens with a double flower.

The leaves being of a foft mucilaginous nature, are boiled and eaten by fome people as a fallad, and are deemed good againft the piles and heat in the fundament.

16 RAPHANUS fativus. *Common Radifh.*
The leaves of this are often boiled as a fallad, and if they be young and tender, they eat very agreeably.

17 SALVIA fclarea. *Garden Clary. Lin. Sp. pl.* 38.
Horminum Sclarea dictum. *Bauh. Pin.* 228.
This is a biennial, and a native of Italy, but it has poffeffed a place in the Englifh gardens for a long time. The root is fibrous, and fends forth feveral large, whitifh green, oblong, heart - fhaped leaves, which are much wrinkled, ferrated on their edges, and hairy on their furfaces. The ftalks are fquare, hairy, greatly branched, fometimes a little clammy, two or three feet high, and fet at their joints with pairs of leaves like thofe from the root, but fmaller. The branches ftand oppofite, and are terminated by long fpikes of pale blue flowers, placed in whorls, with two whitifh concave, acute pointed leaves under each. The flower-cup is divided into two lips, the upper one ending in three fpiculæ; and the under one in
two.

two. The flower alfo has two lips, the upper one is erect and arched, with one ftyle nearly of the fame length under it, and two ftamina that are fhorter. The lower lip is cut into three fegments. Every part of the plant emits a very ftrong fcent.

The frefh leaves dipped in milk, and then fryed in butter, were formerly ferved up at table as a delicate fallad. Some people too boiled them as a pot-herb. The plant ufed any way is counted excellent againft hyfterical diforders. Of the different parts of it a wine is made, which is a high cordial, and not to be equalled by any other home-made wine. The following is the moft approved Recipe for making it.

To five gallons of cold water, put four pounds of Lifbon fugar, and the whites of three eggs well beaten; boil thefe together gently about an hour, then fkim the liquor, and when it is almoft cold, add of the fmall Clary leaves and the tops in bloffom, one peck, and alfo half a pint of ale yeaft. This done, put the whole into a veffel, and ftir it twice a day till it has done working, then ftop it clofe for eight weeks. After the expiration of this time draw it into a clean veffel, adding to it a pint and half of good Brandy. In two months it will be fit to bottle.

K   18 Spinacia

18 SPINACIA oleracea. *Spinach*. *Lin.*
Sp. pl. 1456.

Lapathum hortenfe five fpinacia femine
fpinofo. *Bauh. Pin.* 114.

Lapathum hortenfe five fpinacia femine
non fpinofo. *Bauh. Pin.* 115.

This is an annual, and is too well known
to require any defcription. What particular
country it is a native of is not certain, but it
is known to have been cultivated in England
more than two hundred years. It hath fa-
gittated leaves and prickly feeds. Linnæus
makes the fmooth-feeded *Spinach* only a variety of this, though it differs as much in
the leaves as in the feeds, thofe of the latter
being egg-fhaped. This laft is the fort
now chiefly cultivated for the kitchen, but
it is a much more tender plant than the
former. *Spinach* is a good fallad for thofe
of a coftive habit of body, as it obtunds the
acrimony of the bowels, and gently relaxes
them.

19 THEA bohea. *Bohea Tea.* *Lin. Syft.*
Nat. 365.

It muft be owned that neither *Tea* nor
Coffee can with ftrict propriety be placed
under any of thefe divifions, becaufe neither
the leaves of the one or the berries of the
other can be truly called efculent; yet to
have entirely omitted them would have
caufed a fort of chafm in the work, by reafon
the

the infusions of both are constantly mingled with our daily food. The leaves of Tea, however, are often eaten by the poorer people after they have been infused; but this is a practice not to be recommended, as they can afford no nourishment, and do certainly much injure the stomach, and the whole nervous system.

The *Bohea* is a shrub that rises about six or eight feet high, and divides into many irregular branches, which are furnished with oval, smooth, glossy, serrated leaves, standing singularly on short footstalks. These are from two to three inches long, one broad, with prominent veins on their under sides, and end in snipped obtuse points. The flowers come out at the bosoms of the leaves, on club-shaped peduncles, more than half an inch long; they consist of six white roundish, concave petals each (two of which are less than the rest) including two or three hundred stamina, surrounding a very short style, crowned with three long, recurved, awl-shaped stigmata. When the flower is fallen, the germen swells to a sort of triangular capsule, composed of three globular cells united, each containing one hard, roundish seed, of a woody texture. The shrub is a native of China and Japan.

20 THEA viridis. *Green Tea. Lin.
Syst. Nat.* 365.

This

This differs in nothing from the former, but that the flower is compofed of nine petals, and the other of but fix.

I have here given the *Thea* as it ftands in the Syftema Naturæ of Linnæus; but tho' this learned Botanift makes two diftinct fpecies of it, yet it is highly probable that all the forts of *Tea* are gathered from one and the fame fpecies, and that the nine petals in the flower is merely accidental. As to the great differences found in the tafte, fmell, and colour of the various kinds, when they are fit for fale, thefe may be occafioned by the different ages of the leaves, the time of collecting, the manner of curing them, by fome vegetable liquid they may be fprinkled with, or the foil and fituation the trees may grow in.

In regard to the medicinal virtues of *Tea*, fome authors make it little better than a poifon, whilft others think it the moft wholefome and falubrious vegetable on earth. A very fuperficial examiner will perceive it to be refrefhing and exhilarating, and that it is excellent for carrying off the effects of a debauch; but notwithftanding thefe good qualities, an immoderate ufe of it will be found to bring on a train of the worft of nervous complaints; and in fome tender conftitutions even a cup or two is feen to throw them into tremors and fpafmodic affections. The green *Teas* feem to bring

on

on thefe bad effects fooner than the boheas, but the finer either fort is, the more its pernicious confequences are to be dreaded.

21 URTICA dioica. *Common Stinging Nettle. Lin. Sp. pl.* 1396.
Urtica urens maxima. *Bauh. Pin.* 232.

It is a common practice now, among the ordinary people, to gather the leaves and young fhoots of the common *Stinging Nettle* in the fpring, and boil them for a fallad; and if the better fort were to follow their example, they might often find a benefit by it. Thefe leaves are not unpleafant to the palate, are an excellent antifcorbutic, and powerful againft all cutaneous eruptions. I have known fome inftances where they have been ufed in this manner once a day, by thofe all covered with blotches, and in a month's time their fkins have become per-fectly fmooth, and free from any deformity. The roots are in high efteem for ftopping the fpitting of blood, and bloody urine. Thefe are very diuretic, and a decoction of them drank frequently is faid to be fo powerful, as to break the ftone in the bladder.

K 3   SECT

§ E C T. III.

Pot-herbs.

1 APIUM graveolens. *Celery.* See
the firſt *Chap.*

2 Apium petroſelinum. *Parſley.* See
Ditto.

3 Allium porrum. *Leeks.*

4. Braſſica oleracea. *Cabbages.* See the
former *Sect.*

5 Beta vulgaris alba. *White Beet.*

6 Chrithmum maritimum. *Rock Samphire.*

7 Hyſſopus officinalis. *Common Hyſſop.*

8 Oxalis acetoſella. *Wood Sorrel.* See
the firſt *Sect.*

9 Ocymum baſilicum. *Sweet-ſcented Baſil.*

10 Origanum majorana. *Common Marjoram.*

———— *majorana tenuifolia.* Fine-
leaved Sweet Marjoram.

11 Origanum heracleoticum. *Winter Sweet
Marjoram.*

12 Origanum onites. *Pot Marjoram.*

13 Picris echioides. *Common Oxtongue.*

14 Roſmarinus officinalis. *Common Roſemary.*

— Roſma-

— *Rosmarinus hortensis.*   Garden Rose-
mary.
15 Salvia officinalis. *Green and Red Sage.*
———— *minor.* Tea Sage.
16 Satureja hortensis. *Summer Savory.*
17 Satureja montana. *Winter Savory.*
18 Scandix cerefolium. *Common Chervil.*⎫
19 Scandix odorata, *Sweet Cicely.*   ⎬
    See the first *Sect.*   ⎭
20 Sonchus oleraceus. *Common Sow-thistle.*
21 Thymus vulgaris. *Common Thyme.*
22 Thymus mastichinus. *Mastick Thyme.*

    3 ALLIUM porrum, *Leeks.* *Lin. Sp.*
pl. 423.
    Porrum sativum latifolium. *Bauh. Pin.*
72.
    This plant has been so long cultivated
that its native place of growth cannot be
traced.   It is undoubtedly the same as that
mentioned in the xi Chap. of Numbers,
where it is said the Israelites longed for
Leeks in conjunction with Onions.   The
leaves are much of the same nature as those
of the latter, and they are yet a constant
dish at the tables of the Egyptians, who
chop them small and then eat them with
their meat.   They are in great esteem too
with the Welsh, and their use as a pot-herb
with the English is well known.

    5 The *Beta alba* is only a variety of the
red

red *Beet*, and is but rarely ufed now to
what it was formerly. It is generally mixed with favory herbs, it being too infipid to
impart much flavour of itfelf. Both the
juice and powder of the root are good to
excite fneezing, and will bring away a confiderable quantity of mucus.

6 CRITHMUM maritimum. *Rock Samphire. Lin. Sp. pl.* 354

Crithmum, Fœniculum maritimum minus. *Bauh. Pin.* 288.

This is a low perennial plant, and grows
upon rocks by the fea in feveral parts of
England. It has a fpicy, aromatic flavour,
which induces the poor people to ufe it as
a Pot-herb. It is alfo gathered and fold
about for the purpofe of pickling, and it is
in great efteem when thus managed. But it
muft not be underftood here that this is the
Samphire generally pickled in Norfolk, for
that is the *Salicornia europea*, before defcribed. There is another fort of *Samphire*
too, commonly fold about the ftreets and
markets for this *Crithmum*, and is generally
bought by people not fkilled in plants for
the true one. This laft is the *Inula crithmoides*, (Golden Samphire) which, though
it has fome little refemblance to the former, yet it is a plant of a quite different
nature, and far inferior in flavour when
pickled. In order therefore to prevent people

ple being impofed on, I fhall here give a particular defcription of the *Rock Samphire.*

The root of this plant is compofed of feveral tough fibres which penetrate deep into the fiffures of the rocks. It fends forth many green, fucculent ftalks, near half a yard high, ornamented with deep green, winged leaves, compofed of three or five divifions, each of which hath three or five fmall, thick, flefhy lobes, near an inch long, and the bafe of their common pedicle embraces the main ftalk. The flowers are yellowifh, and are produced in circular umbels; they are fmall, confift of five equal petals each, with five ftamina of the fame length, and are fucceeded by feeds like thofe of Fennel, but they are fomewhat larger.

By a proper attention to this defcription the *Crithmum maritimum* may always be diftinguifhed from the *Inula crithmoides,* by fuch as are total ftrangers to the knowledge of plants, for the *Inula* has a flower like that of Flea-bane, and its leaves are linear, except juft at the apex, where they fpread a little, and end in three jags or teeth. The *Crithmum* may be propagated in gardens, provided it be planted on a gravelly foil, and this would be a certain way to avoid the cheat. The medicinal virtues of this plant are thofe of removing obftructions of the vifcera, and urinary paffages.

7 Hyssopus

7 Hyssopus officinalis. *Common Hyſſop.*
Lin. Sp. pl. 796.

Hyſſopus officinarum cærulea five fpicata.
Bauh. Pin. 217,

This plant grows naturally in feveral
parts of Aſia. It is a perennial, and has
been fo long cultivated in gardens, that it
is known by almoſt every one. It is exceeding grateful to the fmell, and ſtands recommended againſt afthmas, coughs, and all
diforders of the breaſt and lungs, whether
boiled in foups or otherwife ufed. There
is a diſtilled water made from it kept in the
fhops, which is deemed a good pectoral.

9 Ocimum baſilicum. *Sweet-fcented*
Baſil. Lin. Sp. pl. 833.

Ocimum caryophyllatum majus. *Bauh.*
Pin. 226.

This is an annual, and a native of Perſia;
fince it has been cultivated in Europe, it
has produced many varieties. The hairy
Baſil, which is that commonly fown in gardens, feems to be no other than one of thefe
varieties, though made a diſtinct fpecies by
Miller and others. This fort rifes near
half a yard high, fending out branches by
pairs in oppofite directions; thefe, and alfo
the main ſtems, are hairy and four fquare.
The leaves are oval, indented about their
edges, and end in a fharp point. The
flowers are of the lip kind, are white, and

5   terminate

terminate the ftalks and branches in long fpikes. The ftamina are four, two longer than the other, and the feeds lie naked at the bottom of the calyx. The whole plant has a ftrong fmell of Cloves.

The French are fo infatuated with the flavour and qualities of it, that its leaves come into the compofition of almoft all their foups and fauces.

10 ORIGANUM majorana. *Summer Sweet Marjoram. Lin. Sp. pl.* 825.

Majorana vulgaris. *Bauh. Pin.* 224.

The natural country of this is not known. It is an annual, and hath oval, obtufe leaves, and almoft round, hairy fpikes. As it lives only one Summer, it will be beft to diftinguifh it by the name of *Summer Sweet Marjoram*, the better to contraft it with the following, which is called *Winter Sweet Marjoram*.

11 ORIGANUM heracleoticum. *Winter Sweet Marjoram. Lin. Sp. pl.* 823.

Origanum heracleoticum, Cunila gallinacea plinii. *Bauh. Pin.* 223.

This is a perennial, and a native of Greece. It hath long fpikes growing in bunches, and flower-leaves as long as the flowercups. It is hardy, and will live through the winter in the open air in our climate; which

which circumſtance is alone ſufficient to di-
ſtinguiſh it from the former.

12 ORIGANUM onites. *Pot Marjoram.*
Lin. Sp. pl. 824.

Majorana major angelica.   *Ger. em.* 664.

This too is a perennial, and has been
found wild in England. In its general habit it is like the *majorana,* but the ſtalks
are more woody, and furniſhed with long
hairs. The leaves are ſmall, heart-ſhaped,
ſharp pointed, on both ſides woolly, ſeldom
ſerrated, and have little or no foot-ſtalks.
The ſpiculæ come out in cluſters, as in the
Common Marjoram, but they are longer,
hairy, and ſtand three upon a common peduncle, the middle one being feſſile, and
all the flowers white.

The uſe of the leaves of all theſe ſpecies
is well known in the kitchen, and therefore
it will be needleſs to ſay any thing about it.
They are all warm aromatics, and are often
preſcribed alone, or in phyſical compoſi-
tions. Half an ounce of the tops of the
majorana, may be infuſed in a pint of boiling
water, and drank occaſionally againſt head-
aches, aſthmas, and catarrhs. The powdered leaves are a good errhine, and are often
uſed for this purpoſe. The *onites* is not
quite ſo gratefully ſcented as the *majorana,*
but it is frequently ordered in baths for
diſorders

diforders in the head, and againſt cutaneous
eruptions. This grows plentifully in Syra-
cuſe, and alſo in ſome parts of Greece.

13 Picris echioides. *Common Ox-tongue.*
Lin. Sp. pl. 1114.

Hieracium echioides capitulis cardui be-
nedicti. *Bauh. Pin.* 128.

This is a native of England, is an annual,
and may be found on the borders of corn-
fields. It ſends forth ſeveral dark green,
oblong oval leaves, having many protuberances on their ſurfaces, and are thickly ſet
with ſtiff hairs. Among the leaves riſes a
round, green, hairy ſtalk, to about two
feet, with a few leaves thereon, and breaking into branches towards the top, which
are furniſhed with ſmall yellow flowers,
ſomewhat like thoſe of the Sow-thiſtle;
theſe are ſucceeded by browniſh long ſeeds,
crowned with down.

The leaves are frequently uſed as a Pot-
herb, and are eſteemed good to relax the
bowels.

14 Rosmarinus officinalis. *Roſemary.*
Lin. Sp. pl. 33.

Roſmarinus ſpontaneus, latiore folio.
Bauh. Pin. 217.

This ſhrub grows in prodigious abundance in the ſouthern parts of Europe. It
is ſo common in gardens as to be known by

2   every

every one. Many people boil the leaves in
milk pottage, to give them an aromatic
flavour. The fprigs too are frequently ftuck
into beef whilft it is roafting, and they communicate to it an excellent relifh. With
the flowers of this plant is made the much
celebrated Hungary water. They are deemed excellent aromatics, and are ufed in all
nervous complaints, that take their rife from
too great cold and moifture in the habit of
body. They abound with a fubtile, pene-
trating oil, which renders them ferviceable
in the jaundice and gout.

15 SALVIA officinalis. *Green and Red
Sage. Lin. Sp. pl.* 34.
   Salvia major. *Bauh. Pin.* 237.
   This is a native of Auftria, and by being
long planted in gardens it comes of two colours, red and green. The fmall *Tea Sage*
too is only a variety of the *officinalis.* This
is the fort that is generally made ufe of for
culinary purpofes, it being the pleafanteft;
but for phyfical intentions, the large kind
ought to be chofen; and in moft cafes the
red fhould have the preference, it being
more corroborating than the green, which
renders it immediately ferviceable in all re-
laxations of the fibres. The ancients had
this plant in the higheft efteem, and perhaps not unjuftly, for it is certainly an excellent vulnerary, and a great ftrengthener

of

of all the internal parts of the body, and particularly the lungs.

16 SATUREJA hortensis. *Lin. Sp. pl.* 795.

The *Summer Savory* is an annual, and a native of France and Italy. It fends forth several flender erect ftalks, near half a yard high, which put forth branches by pairs, and are fet with leaves placed oppofite; thefe are ftiff, a little hairy, and yield a fine aromatick fmell on being rubbed. The moft diftinguifhing mark of this fpecies is, that it has two flowers to every peduncle.

17 SATUREJA montana. *Winter Savory. Lin. Sp. pl.* 794.

This is a perennial, is a more fhrubby plant than the former, and it does not rife fo high. The leaves are of a dark green colour, and fharp pointed. The flowers are fuftained by fingle diverging peduncles, coming at the fides of the branches. The root is woody, and fends forth green leaves all the winter. It is a native of France.

Thefe two plants give place to none of the European aromatics for pleafantnefs of fmell and flavour, nor yet in their ufefulnefs in the kitchen; for befides being ufed as Pot-herbs, they are frequently put into cakes, puddings, faufages, &c. They are

warm

warm and difcuffive, and good againft cru-
dities in the ftomach.

20 Sonchus oleraceus. *Common Sow-
thiftle. Lin. Sp. pl.* 1116.

This is an annual plant, and a very troublefome weed in fields and gardens. It varies fo much in different foils that fome of our
moft difcerning Botanifts have made feveral
diftinct fpecies of it. In fome fituations the
whole plant is fmooth, but in others it is
rough, prickly on the margins and midribs
of the leaves, and alfo on the peduncles and
calyces of the flowers. The ftalks are copioufly ftored with a lactefcent juice.

The leaves have little tafte, except a flight
aftringency, yet they are much ufed in fome
of the northern parts of Europe as a Pot-
herb. They were formerly kept in the
fhops by the names *Sonchi afper et Sonchi
lævis,* but they had not any known virtues
fufficient to fupport their place there. The
whole plant is a favourite food of Rabbits.

21 Thymus vulgaris. *Common Thyme.
Lin. Sp. pl.* 825.

Thymus vulgaris, folio tenuiore et latiore.
Bauh. Pin. 219.

The *Thymus vulgaris* grows wild on the
mountainous parts of France, Spain, and
Italy. This is the broad leaved *Thyme* commonly

monly cultivated in gardens, and therefore is well known.

22 THYMUS maſtichinus. *Maſtick Thyme.* *Lin. Sp: pl.* 827.

Sampſucus, five Marum maſtichen redolens. *Bauh. Pin:* 224.

This plant grows ſpontaneouſly in Spain. It is a perennial, of a tenderer nature than the former, and differs much from it in its general habit, which induced Miller to place it among his Satureja. The ſtalks riſe about half a yard high, breaking into ſlender, woody branches, which are covered with a brown bark, and ſet with leaves like thoſe of the *vulgaris* in ſhape, but they are rather larger. The flowers come out in whorls at the tops of the branches, and are ſurrounded with a greyiſh wool; they are white, with briſtly, denticulated cups.

Both theſe plants are fine aromaticks, and are uſed in the kitchen for the ſame purpoſes as the *Savories.* The dried leaves and tops of the *maſtichinus* are ſaid to be powerful againſt an immoderate flow of the menſes. A dram of the powder in a glaſs of red wine is a doſe.

L   C H A P.

C H A P. IV.

ESCULENT FLOWERS.

1 CALENDULA officinalis. *Common Marigold.*
2 Caltha paluftris. *Marſh Marigold.*
3 Capparis fpinofa. *Caper Buſh.*
4 Carthamus tinctorius. *Safflower.*
5 Carlina acaulis. *Dwarf Carline Thiſtle.*
6 Cynara cardunculus. *Cardoon.*
7 Cynara fcolymus. *Green or French Ar-tichoke.*
———- *hortenſis.* Globe Artichoke.
8 Cercis filiquaftrum. *Common Judas-tree.*
9 Helianthus annuus. *Annual Sun-flower.*
10 Onopordum acanthium. *Cotton Thiſtle.*
11 Tropæolum majus. *Indian Creſs, or Naſturtium.*
12 Tropæolum minus. *Smaller Indian Creſs.*

1 CALENDULA officinalis. *Common Ma-rigold. Lin. Sp. pl.* 1304.
Caltha vulgaris. *Bauh. Pin.* 275.
This is fo very common in gardens as to make it univerfally known. It is a native of Spain. The flowers gathered and then dried were formerly in high efteem among
houfe-

houfe-keepers to boil in foups and pottage.
They are deemed cordial, and a refresher
of the animal spirits. There are many varieties of this plant raifed in gardens, more
for ornament than ufe.

2 CALTHA paluftris. *Marfh Marigold.*
Lin. Sp. pl. 784.
Caltha paluftris, flore fimplici. *Bauh.*
Pin. 276.
The *Caltha paluftris* is a perennial, and
the only plant yet known of the genus. It
is very common in our meadows, where it
fends forth many large, roundifh heart-fhaped
leaves, flightly crenated on their edges, a-
mong which rife round, hollow, green
ftalks, dividing into three or four branches
towards their top, and having a feffile leave
at each divifion. The flower is compofed of
five large oval, concave yellow petals, furrounding many flender ftamina, and feveral
oblong, compreffed germina, or feed-buds,
which become as many pointed capfules,
containing feveral roundifh feeds. It flowers early in the fpring, when its yellow
flowers are a great ornament to the meadows. There is a variety of it in gardens
with a double flower.

The flower-buds of this plant are by
many people pickled as Capers, for which
they are a good fubftitute.

3 CApparis fpinofa. *Caper Bufh. Lin. Sp. pl.* 720.

Capparis fpinofa, fructu minore, folio rotundo. *Bauh. Pin.* 480. _

This is a low fhrubby plant, and a native of Italy. It fends forth woody ftalks, which divide into many flender branches, under each of which are placed two fhort crooked fpines, and between thefe and the branches come out round, fmooth leaves, fingly upon fhort foot-ftalks. At the infertions of the branches iffue the flowers; thefe are white, and compofed of five roundifh concave petals each, furrounding a great many flender ftamina, and one ftyle longer than the ftamina, fitting upon an oval germen, which turns to a capfule filled with kidney-fhaped feeds. The flower when fully expanded looks like a fingle white Rofe.

The buds of thefe flowers are pickled, and annually fent into England, and other places, by the name of *Capers.* They are faid to excite the appetite, promote digeftion, and to help obftructions of the liver and fpleen; but it is probable thefe valuable qualities proceed more from the ingredients they are pickled in, than from the *Capers* themfelves.

4 Carthamus tinctorius. *Safflower. Lin. Sp. pl.* 1162.

Cnicus

Cnicus fativus, five Carthamus officinarum. *Bauh. Pin.* 378.

This is an annual plant, and a native of Egypt. It fends up a ftiff woody ftalk, to two feet or more high, breaking into many branches, which are furnifhed with oval, fharp-pointed, feffile leaves, flightly jagged on the edges, and each jag ending with a fharp fpine. The flowers terminate the branches in large, fcaly heads. The fcales are flat, broad at their bafe, and taper to a point, where they terminate in a fharp fpine. The florets are numerous, funnel-fhaped, of a fine faffron colour, and ftand up above the fcales of the empalement near an inch. They are all hermaphrodite, and are fucceeded by white, fmooth, oblong feeds, near as large as wheat.

Formerly the common people ufed to put the dried florets into their puddings, I fuppofe more to give them a colour, than for any good flavour the flowers communicated; when this was done in large quantities, the puddings proved purgative, whereby the practice is now quite laid afide.

This plant is cultivated in great abundance in Germany, whence the other parts of Europe are fupplied with the flowers, which form a great article of trade, they being ufed in dying and painting. If they be neatly dried, it is difficult to diftinguifh them from Saffron, but by the fmell. The

feeds

feeds are kept in the fhops, and have been in repute as a good cathartic, but their operation is flow and not always certain.

5 CARLINA acaulis. *Dwarf Carline Thiftle. Lin. Sp. pl.* 1160.

Carlina acaulos, magno flore albo. *Bauh. Pin.* 380.

This *Thiftle* grows on the mountainous parts of Italy and Germany. It hath many large whitifh green, finuated leaves, laying on the ground, which are fet with fmall fharp fpines round about their edges. In the centre of thefe comes a large flower-bud, without any ftalk, but is furrounded with long, prickly, jagged leaves, adhering to its bafe. The flower is compofed of white, hermaphrodite florets, which are fucceeded by roundifh, white feeds, crowned with a branched, feathery down.

The central part of the flower is boiled and eaten the fame as Artichoke bottoms. The root is kept in the fhops; it is of a brown rufty colour, about an inch thick, very porous, fo that when cut it appears as if worm-eaten. It has a ftrong fmell, and a bitterifh tafte, mixed with a flight degree of aromatic. It was in high efteem among the ancients as a diaphoretic.

6 The *Cynara cardunculus,* or *Cardoon,* was defcribed in the fecond Chapter, among
the

the stalks; I have given it a place here upon the authority of some travellers, who have assured me that the heads are also eaten, but I doubt they mistook the species.

7 CYNARA scolymus. *Green or French Artichoke. Lin. Sp. pl.* 1159.
Cynara sylvestris latifolia. *Bauh. Pin.* 384.

This grows wild in the fields of Italy, and Linnæus makes the *hortensis* only a variety of it. The latter is that sort which is now chiefly cultivated, by reason the bottoms are more fleshy, and much better tasted than those of the *scolymus*. The use they are put to in the kitchen is so well known, that to say any thing about it will be quite unnecessary.

8 CERCIS siliquastrum. *Common Judas-tree. Lin. Sp. pl.* 534.
Siliqua sylvestris rotundifolia. *Bauh. Pin.* 402.

The *Common Judas-tree* grows in France, Spain, and Italy. It rises with a straight trunk, covered with a reddish bark, to the height of twelve or fourteen feet, dividing towards the top into many irregular branches, furnished with roundish heart-shaped, smooth leaves, having long footstalks. The flowers come out in clusters from all sides of the branches, and sometimes even from

L 4. the

the trunk itſelf; they are of a bright purple colour, ſtand upon ſhort peduncles, have five petals each, reſembling a pea-bloom, and ten diſtinct ſtamina, four of which are longer than the reſt, and ſurround a long, ſlender germen, which becomes a long flat pod, having one cell, containing many roundiſh ſeeds.

The flowers have a ſharp, acid flavour, and are not only mixed with ſallads to render them more grateful, but are alſo pickled in the bud, in the manner of Capers.

The wood of this tree is hard, and beautifully veined with black and green. It will take a fine poliſh, and on that account is converted to many fanciful uſes.

9 HELIANTHUS annuus. *Annual Sun-flower. Lin. Sp. pl.* 1276.

Helenium indicum maximum. *Bauh. Pin.* 276.

This is a native of America, but is now ſown in almoſt every garden in England, on account of its bold, large, yellow flowers, which make a fine appearance in the autumn. The bottoms of theſe flowers are very fleſhy, and many people dreſs and eat them, as they do thoſe of the Artichoke.

The ſeeds of this plant are copiouſly ſtored with oil, which may be eaſily expreſſed, and is not inferior to that drawn from Olives. The ſeeds have as agreeable

a flavour

a flavour as Almonds, and are excellent food for domeſtic poultry.

10 The *Onopordum acanthium*, or *Cotton Thiſtle*, has been deſcribed in a former Chapter; it ſtands here by reaſon the bottoms of its flowers are eaten in the manner of thoſe abovementioned.

11 TROPÆOLUM majus. *Indian Creſs. Lin. Sp. pl.* 490.

Acriviola maxima odorata. *Boerh. lugdb.* I. *p.* 244.

This is a native of Peru, and an annual. It hath weak trailing ſtalks, which are furniſhed with ſmooth, greyiſh green, almoſt circular leaves, ſupported on long footſtalks, inſerted into their centre. The flowers are produced from the ſides of the ſtalks; they are in ſome plants of a pale yellow, in others of a deep orange colour, and are of a ſingular ſtructure, being compoſed of five petals, the upper two of which are broad, the three under ones narrow, their baſes joined together, and lengthened into a ſpur above an inch long. They include eight declining, awl-ſhaped ſtamina, and a roundiſh, ſtreaked germen, ſupporting one erect ſtyle, crowned by an acute trifid ſtigma. The germen becomes a furrowed berry, divided into three lobes, each including one ſtriated ſeed.

12 TRO-

12 TROPÆOLUM minus. *Smaller Indian Cress. Lin. Sp. pl.* 490.

Nafturtium Indicum. *Ger.* 196.

This is a native of Peru and other parts of South America. It differs from the former in the leaves being entire, the other having five obfolete lobes ; the petals of the flower of this are fharp-pointed and briftly, thofe of the *majus* are obtufe. There is a variety of this fort with double flowers. Thefe plants being very ornamental, are now annually fown in moft gardens, for they flower a long time, and make a beautiful appearance.

The flowers have a fragrant fmell, and a fharp pungent tafte, like that of Garden Creffes. In France they are not only ufed to garnifh difhes, but are mixed with Lettuce and other cold fallads, and are efteemed both pleafant and wholefome. The berries have a warm fpicy flavour, and make an excellent pickle.

C H A P.

CHAP. V.

ESCULENT BERRIES.

SECT. I.

Indigenous, or native Berries *.

1 ARBUTUS uva urfi. *Bearberry.*
2 Arbutus alpina. *Mountain Straw-berry.*
3 Arbutus unedo. *Common Strawberry-tree.*
4 Berberis vulgaris. *Common Berberry.*
5 Cratægus aira. *White Beam-tree.*
6 Cratægus torminalis. *Maple-leaved Ser-vice or Sorb.*
7 Fragaria vefca, vel fylveftris. *Wood Strawberry.*
———— *northumbrienfis.* Northumber-land Strawberry.
———— *imperialis.* Royal Wood Straw-berry.
———— *granulofa.* Minion Wood Straw-berry.
8 Fragaria viridis vel pratenfis. *Swedifh Green Strawberry.*

* A *Berry* is defined by Linnæus to be a pulpy *feed-veffel*, without a valve, and inciofing feveral feeds, which have no other covering.

9 Fragaria

9 Fragaria mofchata. *Hautboy Strawberry.*
—— *mofchata rubra.* Red-bloſſomed
Strawberry.
—— *mofchata hermaphrodita.* Royal
Hautboy.

10 Fragaria chinenſis. *Chineſe Strawberry.*

11 Fragaria virginiana. *Virginian Scarlet
Strawberry.*
—— *virginiana coccinea.* Virginian
ſcarlet-bloſſomed Strawberry.
—— *virginiana campeſtris.* Wild
Virginian Strawberry.

12 Fragaria chiloenſis. *Chili Strawberry.*
—— *chiloenſis devonenſis.* Devonſhire
Strawberry.

13 Juniperus communis. *Common, or Engliſh Juniper.*
—— *arbor.* Swediſh Juniper.

14 Ribes rubrum vel album. *Red and
White Currants.*

15 Ribes nigrum. *Black Currants.*

16 Ribes groſſularia. *Gooſeberries.*

17 Roſa canina. *Dog's Roſe, or Hep-buſh.*

18 Rubus idæus. *Raſpberry.*
—— *idæus albus.* White Raſpberry.
—— *idæus lævis.* Smooth - ſtalked
Raſpberry.

19 Rubus cæſius. *Dewberry.*

20 Rubus fruticoſus. *Common Bramble.*

21 Rubus chamæmorus. *Cloudberry.*

22 Rubus arcticus. *Shrubby Strawberry.*

23 Vaccinium

23 Vaccinium myrtillus. *Blackworts, or Bilberry.*

24 Vaccinium vitis idæa. *Redworts.*

25 Vaccinium oxycoccos. *Cranberry.*

1 ARBUTUS uva urfi. *Bearberry. Lin. Sp. pl. 566.*

Radix idæa putata et uva urfi. *Bauh. Hift. I. p. 524.*

This plant grows naturally in the northern parts of England. It is a fmall fhrub, rifing little more than a foot high, breaking into many branches, which are clofely fet with fmooth, thick, oval leaves, entire on their margins. The flowers are produced in fmall bunches, near the extremities of the branches; they have an obtufe, quinquefid *, purple calyx, furrounding a pitcher-fhaped, white petal, cut at the brim into five teeth, which roll backwards, and contain ten awl-fhaped ftamina, and a cylindrical ftyle. The germen is roundifh, and becomes an oval, or globular berry, having five cells, filled with fmall, hard feeds.

2 ARBUTUS alpina. *Mountain Straw-berry. Lin. Sp. pl. 566.*

Vitis idæa foliis oblongis albicantibus. *Bauh. Pin. 470.*

This grows upon the Alps, alfo in Lap-land and Siberia, and has been found too in fome parts of England. The branches are

* Cut into five parts.

flender,

flender, and trail upon the ground; thefe are furnifhed with oblong, rough, ferrated, whitifh green leaves. The flowers are produced from the wings of the leaves, upon long, flender peduncles, and are fucceeded by berries about the fize of black Cherries; thefe are green at firft, red afterwards, and black when ripe.

3 ARBUTUS unedo. *Common Strawberry-tree. Lin. Sp. pl.* 566.

Arbutus folio ferrato. *Baub. Pin.* 460.

This tree grows very plentifully in the woods in Ireland, but is common now in the Englifh gardens, being a very ornamental plant, it having ripe fruit and flowers upon it at the fame time; for the flowers blow in the autumn, and the fruit that fucceed them hang till the next autumn before they are ripe, when a frefh fet of flowers puts forth, and fo on. The fruit have an auftere, four flavour, yet they are eaten by the Irifh, who are very fond of acids, and are fold in their markets. There are feveral varieties of this fpecies, but thofe moft commonly cultivated are the red flowered, and the double flowered. The fruit of the two firft forts are not of a delicate flavour, yet they are eaten by the inhabitants where the plants grow naturally.

The leaves of thefe plants are all aftrin-gent, and thofe of the *uva urfi* have been

said

said to do wonders in the gravel. For this purpose half a dram of the powder is ordered in any convenient vehicle once a day.

4 BERBERIS vulgaris. *Common Berberry.* *Lin. Sp. pl.* 471.

Berberis dumetorum. *Bauh. Pin.* 454.

This is common in hedges in many parts of England, and sends forth several stalks eight or ten feet high; these run into numerous branches, covered with a whitish bark, and are armed with short spines, which generally come out by three at a place. The leaves are egg-shaped, obtuse, finely serrated on the edges, and when chewed have an acid, astringent taste. The flowers are yellow, and are produced in long bunches in the manner of Currants, each consisting of six roundish, concave petals, having two glands fixed to their base, and include six stamina, with two summits fastened on each side their apex. The germen is cylindrical, and turns to an obtuse, umbilicated berry, of one cell, enclosing two cylindrical seeds. There is a variety of this shrub without any seeds in the berries.

These berries have an agreeable acid taste, and on that account they are boiled in soups to give them a tart flavour. They are also pickled for the purpose of ornamenting dishes. In medicine they are chiefly used in conserve, and in this form they are cooling

ing and aftringent, good to quench thirft,
fortify the ftomach, and ftop diarrhæas and
dyfenteries.

5 CRATÆGUS aira. *White Beam-tree.*
Lin. Sp. pl. 681.

Alni effigie, lanato folio major. *Bauh.*
Pin. 452.

This grows wild in Kent, and fome other
parts of England. It arrives to the height
of thirty feet or more, with a large trunk,
that divides upwards into many branches,
which fpread in the form of a pyramid, the
young twigs being covered with a brown
bark, fprinkled with a mealy down, and
garnifhed with oval leaves, of a light green
colour on their upper fide, white on their
under, unequally ferrated on their edges,
and having many prominent veins running
from the midrib to the border. The flowers come out in bunches at the extremities
of the branches, having mealy peduncles
and empalements; the latter are cut into
five obtufe fegments, fuftaining five fhort,
concave, white petals, which fpread open,
and furround many ftamina, and two ftyles.
When the flower falls, the germen becomes
a roundifh berry, enclofing two oblong hard
feeds.

6 CRATÆGUS torminalis. *Maple-leaved*
Service-tree. *Lin. Sp. pl.* 681.

Sorbus

Sorbus torminalis et Cratægus theophrasti.
Bauh. Hist. I. *p.* 63.

This grows in woods in some parts of
England; it is a taller tree than the former,
and the young branches are covered with a
purplish bark. The leaves are of a bright
green on the upper side, a little woolly underneath, are three or four inches broad,
and shaped like those of the Maple. The
flowers come out in large bunches near the
ends of the branches; they are like those
of the Pear-tree, but smaller, and are succeeded by fruit resembling large haws.

The fruit of both these species are rough
and austere when fresh off the trees, but if
kept in the manner of Medlars, they obtain
an agreeable acid flavour. Those of the
torminalis are annually sold in the London
markets in autumn.

7 FRAGARIA vesca. *Wood Strawberry.*
Lin. Sp. pl. 708.

Fragaria vulgaris. *Bauh. Pin.* 326.

Mr. Weston has published a catalogue of
six distinct species, and sixty varieties of
Strawberries, but Linnæus includes them
all under the *vesca*, or Wood Strawberry, of
which he has two varieties, viz. the *pratensis*, which is the *viridis* of Weston, and
the *chiloensis.*—Besides these two Mr. Weston has the *moschata*, the *chinensis*, and the
virginiana, which, with the *vesca*, make six

diftinct fpecies. I have inferted thefe fix
fpecies, with fuch varieties of them, as Mr.
Wefton judges moft valuable for their fruit,
and fhall here give a fhort defcription of
each variety in his own words.

" The *northumbrienfis* (mentioned by
Wallis in his Nat. Hift.) is a variety of the
common Wood Strawberry, growing naturally in that country; the fruit is red, the
fhape conic, of the fize of a fmall nutmeg,
finer, he fays, than the garden kind. They
grow about twenty miles weft of Newcaftle,
at the beginning of Gofton-burn, on the
north fide, and on the ftrand of the brook
at Hatfield, by the path to Simon-burn.

The *imperialis* is a curious Strawberry,
which was raifed from the Alpine, impregnated by the Wood Strawberry. It was
procured from Lincolnfhire, and it produces
abundance of fruit, which in fize, colour,
and flavour, refemble the Alpine.

The *granulofa* is a fine Strawberry, which,
as well as feveral other varieties, have lately
been obtained from feed, by Monfieur Du-
chefne, one of the moft ingenious Botanifts
of the prefent age.

8 FRAGARIA viridis. *Wefton's Botanicus
Univerfalis. Vol. ii. p. 325.*

It grows plentifully on the hills, and in
the open fields in Sweden, and is later than
the Wood Strawberry. The flefh is firm,
green,

green, and refembles the Nectarine in flavour. The plant is rather low, and remarkable for loofing all its leaves in the winter.

9 FRAGARIA mofchata. *Wefton's Botanicus Univerfalis. Vol.* ii. *p.* 325.

———— *mofchata rubra.* This beautiful variety flowered with me laft year, and is perhaps the fame as that entitled by Jonequet, in his Index Onomafticus, page 49, *Fragaria Americana hirfuta, flore rubro odore mofchi.*

———— *mofchata hermaphrodita.* This moft curious Strawberry has been lately raifed from feeds, and merits the preference on account of its being hermaphrodite. There are alfo feveral other varieties of the Hautboy, differing in fhape, colour and tafte.

10 FRAGARIA chinenfis. *Wefton's Botanicus Univerfalis. Vol.* ii. *p.* 325.

The feeds of this have been lately brought to Europe; and the plant is now firft raifed in the royal gardens at Trianon, but as yet it is too young to produce fruit.

11 FRAGARIA virginiana. *Wefton's Botanicus Univerfalis. Vol.* ii. *p.* 326.

———— *virginiana coccinea.* This un-

commen

common variety is faid to be growing at Worb, in Switzerland.

———— *virginiana campeſtris.* This was introduced into England by Mr. Young, Botaniſt to his Majeſty, in 1772.

12 FRAGARIA chiloenſis. *Weſton's Botanicus Univerſalis. Vol.* ii. *p.* 326.

———— *chiloenſis devonenſis.* This was lately brought from abroad by a curious gentleman, in Devonſhire, and firſt cultivated in the gardens there. The fruit is very large, firm and high-flavoured, in colour nearly approaching to that of the Scarlet Strawberry, and what is extremely ſingular, it bears beſt without any cultivation, and let run wild, except taking off a few of the runners when in bloom. Nor does it want to be renewed or tranſplanted like all the other Strawberries, but will continue fruitful for many years in the ſame bed."

No Engliſh fruit can ſtand in competition with Strawberries for wholeſome and ſalubrious qualities; even their ſmell is refreſhing to the ſpirits, and eaten any way they are delicious. Nor is an immoderate uſe of them attended with any bad conſequences, as is the caſe with Plums, and many other ſorts of fruit. They abate heat, quench thirſt, promote urine, and are gently laxative. Thoſe afflicted with the gout have found great ben. fit by eating plentifully of

them;

them; and Hoffman fays, he has known
confumptions cured by them. So wholefome and pleafant a fruit can never be too
generally cultivated.

The leaves of thefe plants are moderately
aftringent, and are often ufed in gargarifms
for fore mouths, quinfies, and ulcers in the
throat.

13 JUNIPERUS communis. *Common Juniper*. *Lin. Sp. pl.* 1470.

Juniperus vulgaris fruticofa. *Bauh. Pin.*
488.

The common *Juniper* grows naturally in
feveral parts of England, but is frequently
planted in gardens, which makes it generally
known. The *Juniperus arbor*, or Swedifh
Juniper, is only a variety of it, though it
grows three times as large.

The Swedes make an extract from the
berries of this tree, which they generally
eat with their bread for breakfaft, as we do
butter. Of the tops of the branches of the
Canadian pitch-tree, and Juniper-berries, a
very good and wholefome wine is prepared.

The ancient phyficians entertained an opinion of the extraordinary qualities of this
tree, that fell little fhort of enthufiafm, and
held themfelves capable of curing almoft
every difeafe incident to the human body,
by fome preparation or other of the *Juniper*,
as any one may fee by cafting his eye into

Gerard,

Gerard, Parkinſon, and others. Though
it is evident they greatly magnified its virtues, yet it is alſo certain that it is a tree of
vaſt utility, as there are ſeveral excellent
preparations from it ſtill in uſe; as the rob,
the eſſential oil, and compound water of the
berries. The oil is very bitter, and will
effectually kill worms. The wood and roſin
are uſed, but the berries are ſuppoſed to
contain the whole virtues of the tree; they
fortify the ſtomach, diſſipate wind in the
bowels, and are ſaid to be effectual againſt
epidemical infections. The growth of theſe
trees ought to be encouraged near dwellings,
as the perſpirable matter that flows from
them is certainly a means of purifying the
air, rendering it balſamic, and conſequently
ſalubrious.

　　14 RIBES rubrum vel album. *Lin. Sp.*
pl. 290.
　　Ribes vulgare acidum. *Bauh. Hiſt.* ii.
p. 97.
　　The *Red Currant* grows naturally in
Sweden, and other northern parts of Europe.
The white Currant is only a variety of it,
and was at firſt accidentally produced
by culture. The fruit of this ſhrub are
known by all to be grateful and cooling to
the ſtomach, to quench thirſt, and that they
may be eaten in conſiderable quantities
without danger. The jelly made with ſugar
　　5　　　　　　　　　　　　　　　and

and the juice of this fruit is ufed many ways at table, and is an excellent medicine for cooling the mouth in fevers.

15 RIBES nigrum. *Black Currant. Lin. Sp. pl.* 291.
Groffularia non fpinofa, fructu nigro. *Bauh. Pin.* 455.
This is a native of England, and is common by the edges of brooks, and in moift woods. The berries are commonly called *Quinancy-berries*, from their fuppofed excellence againft the Quinfy. A Rob is' made of them, which is frequently adminiftered for this diforder. Though they are rough and aftringent, yet frefh off the bufh they prove laxative to many conftitutions, and are often eaten for this purpofe.

16 RIBES groffularia. *Goofeberry. Lin. Sp. pl.* 291.
The *Goofeberry* is a native of the north of Europe. There is fcarce any fruit capable of more improvement than this, nor any attended with lefs expence in the cultivation. To enumerate its varieties would be quite tedious, and almoft impoffible, for catalogues have been publifhed of near a hundred, and every year is producing new ones. Some of thefe varieties are equal in flavour to the moft efteemed wall-fruit.

M 4   17 ROSA

17 Rosa canina. *Dogs Rose.* *Lin. Sp. pl.* 704.

Rosa sylvestris vulgaris, flore odorato in-carnato. *Bauh. Pin.* 483.

The *Dogs Rose* is known to every one, by being so common in woods and hedges. These berries when mellowed by the frost have a very grateful acid flavour, which tempt many to eat them crude from the bush; but this is a bad practice, for the seeds are surrounded by a hairy, bristly substance, which if swallowed with the pulp, will, by pricking and vellicating the coats of the stomach and bowels, many times occasions sickness, and an itching uneasiness in the fundament. To avoid this therefore the pulp should be carefully cleansed of this matter before eaten. There is a conserve of Heps kept in the shops, which is deemed good in consumptions and disorders of the breast; and in coughs, from tickling defluxions of rheum.

Notwithstanding what has been observed of the bad effects often attending the swallowing that bristly matter found in Heps, yet it is probable this substance might be turned to advantage in some disorders, if judiciously managed; for it is nearly of the same nature to the celebrated *Cow-itch,* so much in use among the Indians for killing of worms, and which they scrape off the pods of the *Dolichos urens.* Their manner

of

of giving the *Cow-itch*, is to mix a fmall
quantity of it with fyrup or honey, and then
eat it for two or three fucceeding mornings
fafting; this done they take a dofe of Rhubarb, and if there be worms it feldom fails
to bring them away. It is plain from this
that the creatures receive their death by being
ftung and pricked with the *Cow-itch*; and
if this matter were given in the fame manner,
why fhould it not have the fame effect? as it
is much of the fame prickly, ftinging nature.

18 Rubus idæus. *Rafpberry. Lin. Sp.
pl.* 706.
Rubus idæus fpinofus. *Bauh. Pin.* 479.
This is a native of our woods, whence it
was tranfplanted into gardens, where it has
produced fome varieties, among which is
that with white fruit. Thefe fruits have a
fine fragrance, but are inferior to the Strawberry in flavour. A fyrup is prepared from
them, and kept in the fhops; this is prefcribed in gargarifms, and is accounted good
againft vomiting, and laxity of the bowels.

19 Rubus cæfius. *Dewberry. Lin. Sp.
pl.* 706.
Rubus repens, fructu cæfio. *Bauh. Pin.*
479.
This too is common in our woods, and
has fome refemblance to the common Bramble, but the ftalks are more weak and trailing,

ing, and the whole plant is smaller. It
may easily be diftinguifhed from the common Bramble by its fruit being not so large,
compofed of fewer knobs, and their being
covered with a blue flue, like plums. Thefe
fruit have a very pleafant tafte, and fteeped
in red wine are faid to communicate to it a
moft agreeable flavour.

20 RUBUS fruticofus. *Common Bramble.*
Lin. Sp. pl. 707.
Rubus vulgaris, five Rubus fructu nigro.
Bauh. Pin. 479.
The *Bramble* is fo common that it is
known by every child. There are two varieties of it; one with white fruit, and another with a white double flower. The berries of this fhrub are eaten in abundance by
children, but they often receive a deal of
hurt from them; they being apt to fwell
the ftomach, and caufe great ficknefs, if eat-
en in any large quantities.

21 RUBUS chamæmorus. *The Cloudberry. Lin. Sp. pl.* 708.
Chamæ Rubus foliis ribes. *Bauh. Pin.*
480.
This grows wild in Weftmoreland, and
fome other places in England; but in Norway and Sweden it is very plentiful. It is
a fmall perennial plant, feldom rifing more
than eight inches high. The ftalks are weak,
                                    without

without fpines, and moftly garnifhed with two or three leaves, nearly the fhape of thofe of the Currant. Each ftalk is terminated by one purplifh flower, which is fucceeded by a blackifh berry, fomewhat refembling that of the Dew-berry.

Thefe berries form an article of trade among the Norwegians, for they collect great quantities of them, and fend them annually to the capital of Sweden, where they are ferved up in deferts at table. They are a favourite fruit too with the Laplanders, who, that they may have recourfe to them at all feafons, bury them in the fnow, and thus keep them from one year to another.

The plant is male and female in diftinct ftems, and is perhaps one of the moft fingular in nature, for the late Dr. Solander obferved, that the male was joined to the female under ground, where they were united into one plant by their creeping roots.

22 Rubus arcticus. *Shrubby Strawberry.* *Sp. pl.* 708.

This is a fmall perennial plant, and grows on the moffy-bogs of Norway, Sweden, and Siberia. It fends forth a few trifoliate leaves, like thofe of the Strawberry, among which rife the ftalks about four inches high; thefe are without fpines, but are furnifhed with leaves like thofe from the root, and each is terminated with a
purple

purple flower, formed like the reft of the genus, and fucceeded by a red berry, much refembling a Strawberry in fmell and flavour.

Linnæus fays this is the moft excellent of all our European fruits, both for fmell and tafte; its odour is of the moft grateful kind, and as to its flavour, it has fuch a delicate mixture of the fweet and acid, as is not equalled by the beft of our cultivated Strawberries.

23 VACCINIUM myrtillus. *Bilberry*. *Lin. Spl. pl.* 498.

Vitis idæa foliis oblongis crenatis, fructu nigricante. *Bauh. Pin.* 470.

This is a fmall fhrubby plant, and is frequently found in woods and upon heaths. It hath a creeping, woody root, furnifhed with brown flender fibres. It fends forth many crooked, ligneous, angular, flattifh ftalks, which are green upward, where they divide into many irregular branches, furnifhed with oval, ferrated leaves, refembling thofe of the fmall-leaved Myrtle; thefe ftand alternately, have very fhort foot-ftalks, and each has the rudiment of a leaf at its bafe. The flowers come out at the bofoms of the leaves, on fhort peduncles; they confift of one blufh-coloured petal each, fnipped at the brim into five fharp-pointed fegments, and include eight ftamina, tipped with horned summits, fummits, with one ftyle in their centre, crowned with an obtufe ftigma. The fruit are of the fize, fhape, and colour of fmall floes, but have a fort of aperture at their apex, and are divided into four cells, containing a few fmall feeds.

Thefe berries are gathered by the inhabitants where the plants grow, who carry them to market for fale, the buyers making them into tarts and other devices. They are alfo eaten raw with cream and fugar.

24 VACCINIUM Vitis-idæa. *Redworts,* or *Whortle-Berries. Lin. Sp. pl.* 500.

Vitis-idæa foliis fubrotundis non crenatis, baccis rubris. *Bauh. Pin.* 470.

This is exceedingly plentiful in Scotland, and is to be met with on mountainous heaths in the north of England. It is a fmaller plant than the former, and an ever-green. The ftalks rife to about eight inches, are branched, and furnifhed with oval leaves, which are dotted on their underfide. Thefe have fo much the refemblance of thofe of the dwarf-box, that they may eafily be miftaken for the latter at a fmall diftance. The flowers come out in a racemus at the ends of the branches; they hang nodding, are of a pale flefh colour, and when they fall are fucceeded by red berries, about the fize of Currants.

Thefe berries have a more grateful acid
flavour

flavour than the former, and on that account are more eagerly fought after by the
country people, who collect them for the
purpofe of making them into tarts, jellies, &c.

25. VACCINIUM oxycoccos. *Cran-berry.*
Lin. Sp. pl. 500.

Vitis-idæa paluftris. *Bauh. Pin.* 471.

The *Cran-berry* grows upon moorifh bogs
in England, and particularly at Lynn in
Norfolk, and in Lincolnfhire. This is a
more feeble plant than the *Vitis-idæa*, the
branches trailing upon the mofs, and are
not thicker than threads. The leaves are
oval, about the fize of thofe of Thyme, of
a glaucous green on their upper fide, but
white underneath. The flowers come from
the bofoms of the leaves, each ftanding
upon a long peduncle; they are fmall and
red, and are followed by red berries, a little
fpotted.

These berries are preferred to either of the
former. They are collected in large quantities by the country people, who carry them
to market-towns for fale. They are either
made into tarts, or eaten raw with cream
and fugar. If they be a little dried and
then ftopped clofe in bottles, they may be
preferved found from year to year.

S E C T.

S E C T.   II.

Foreign Berries, often raifed in Gardens and Stoves.

1 ANNONA muricata. *Sour Sop.*
2 Annona reticulata. *Cuftard Apple.*
3 Annona fquamofa. *Sweet Sop.*
4 Bromelia ananas. *Pine-apple.*
——— *ananas pyramydato fruftu.* Sugar-loaf Pine-apple.
5 Bromelia karatas. *The Penguin.*
6 Cactus opuntia. *Prickly Pear.*
7 Cactus triangularis. *True Prickly Pear.*
8 Capficum annuum. *Annual Guinea Pepper.*
9 Capficum frutefcens. *Perennial Guinea Pepper.*
10 Carica papaya. *The Papaw or Popo.*
11 Carica pofopofa. *Pear-fhaped Papaw.*
12 Chryfophyllum cainito. *Star-apple.*
13 Chryfophyllum glabrum. *Sapadillo, or Mexican Medlar.*
14 Citrus medica. *Common Citron.*
——— *limon.* Common Lemon.
——— *americana.* The Lime-tree.
15 Citrus aurantium. *Common Orange.*
16 Citrus ducumanus. *Shaddock Orange.*
17 Crateva marmelos. *Bengal Quince.*

18 Diofpyros

18 Diofpyros lotus. *Indian Date Plum.*
19 Diofpyros virginiana. *Pifhamin Plum.*
20 Ficus carica. *Common Fig.*
—— *humilis.* Dwarf Fig.
—— *caprificus.* Hermaphrodite-fruited Fig.
—— *fructu fufco.* Brown-fruited Fig.
—— *fructu violaceo.* Purple-fruited Fig.
21 Ficus Sycomorus. *Sycamore, or Pharaoh's Fig.*
22 Garcinia mangoftana. *Mangofteen.*
23 Morus nigra. *Black-fruited Mulberry.*
24 Morus rubra. *Red-fruited Mulberry.*
25 Morus alba. *White-fruited Mulberry.*
26 Mufa paradifiaca. *Plantain-tree.*
27 Mufa fapientum. *Banana, or fmall-fruited Plantain.*
28 Mefpilus germanica. *Medlar.*
29 Mammea americana. *The Mammee.*
30 Malphigia glabra. *Smooth-leaved Barbadoes Cherry.*
31 Malphigia punicifolia. *Pomegranate-leaved Malphigia.*
32 Paffiflora maliformis. *Apple-fhaped Granadilla.*
33 Paffiflora laurifolia. *Bay-leaved Paffion-flower.*
34 Pfidium pyriferum. *Pear Guava, or Bay Plum.*
35 Pfidium pomiferum. *Apple Guava.*
36 Solanum lycoperficum. *Love Apple.*

37 Solanum.

37 Solanum melongena. *Mad Apple.*
38 Solanum sanctum. *Paleſtine Nightſhade.*
39 Sorbus domeſtica. *True Service-tree.*
40. Trophis americana. *Red-fruited Bucc-*
   phalen.
41 Vitis vinifera. *Common Grapes.*
—— *apyrena.* Corinthian Currants.

1 ANNONA muricata. *Sour Sop. Lin. Sp.*
pl. 756.

Annona foliis oblongo-ovatis nitidis, fructibus ſpinis mollibus tumentibus obſitis.
Browne's Jam. 254.

This tree is a native of America. It riſes to
about twenty feet high, breaking into many
branches, which are but thinly furniſhed
with oblong, ſmooth, lance-ſhaped leaves,
of a ſhining green colour. The calyx conſiſts of three heart-ſhaped, ſharp pointed
leaves, ſurrounding ſix heart-ſhaped petals,
three of which are ſmaller than the reſt.
The ſtamina and ſtyles are numerous, but
exceeding ſhort. The berry is large, oblong
heart-ſhaped, moſtly bent a little near the
apex, of a glaucous green colour, and ſtudded with ſoft pointed ſpines.

This fruit contains a ſoft acid pulp,
which is generally eaten in feveriſh diſorders, and is deemed a good cooler.

2 ANNONA reticulata. *Cuſtard Apple.*
Lin. Sp. pl. 757.

N   ANNONA

ANNONA foliis oblongis undulatis venofis, fructibus areolatis. *Browne's Jam.* 256.

This grows in the fame parts of America as the former, but it is taller, and generally reaches the fize of a large Pear-tree. The leaves are long, narrow, fharp-pointed, of a light green colour, with feveral prominent veins, running tranfverfely. The flower is compofed of fix irregular petals, furrounding many very fhort ftamina and ftyles. The fruit is large, conical, of an orange colour, with a fort of net-work on the furface, and when ripe is full of a fweet, yellowifh pulp, like to cuftard in confiftence, which is of a cooling, refrefhing nature, and much efteemed by the inhabitants.

3 ANNONA fquamofa. *Sweet Sop. Lin. Sp. pl.* 757.

ANNONA foliis oblongo-ovatis undulatis venofis, floribus tripetalis, fructibus mammillatis. *Browne's Jam.* 256.

This is a fmaller tree than either of the former, the leaves are broader, and when rubbed have an agreeable fmell. The fruit is roundifh, fcaly on the furface, of a purplifh colour when ripe, and full of a lufcious fweet pulp, whence the name of *Sweet-Sop.*

4 BROMELIA ananas. *Pine-apple. Lin. Sp. pl.* 408.

Carduus

Carduus brafilianus, foliis aloës. *Bauh. Pin.* 384.

This is a native of New Spain, and is a very extraordinary plant in the manner of its growth and propagation. The root fpreads circularly in the ground, and from its centre fends forth a tough ftalk, which is furrounded at the bottom, and for a confiderable way up, with long, green, ferrated leaves, refembling thofe of a fmall Aloe. At the top of the ftalk ftands the fruit, crowned with a tuft of fine green, fharp-pointed leaves. It has fome refemblance on the outfide to the cone of a Pine, whence the name of *Pine-apple*. The flowers are produced from the protuberances of the fruit, are funnel-fhaped, of a bluifh colour, contain fix awl-fhaped ftamina, which are fhorter than the petals, and one ftyle each. When the flowers are fallen, the fruit enlarges, and becomes a flefhy, knobbed berry, plentifully ftored with an exquifite flavoured juice. The feeds are lodged in the knobs; they are very fmall, and nearly kidney-fhaped. A little before the fruit is ripe, there fhoot from the ftalk at the bottom of the berry three or four fuckers, which if taken off and planted, will in about fourteen months produce fruit. The tuft of leaves alfo, taken from the top of the berry, if planted, will do the fame, but not in fo fhort a time. There are feveral

varieties

varieties of the *Pine-apple*, but the moſt
eſteemed ones are the Queen-pine, the
Sugar-loaf, and the Surinam.

This fruit may juſtly challenge all others,
except the *Mangoſteen*, for the delicate and
agreeable variety of its flavour. It ſhould
not ſtand till it is over ripe, and ought to be
eaten almoſt as ſoon as cut. It has been
introduced into England but a little above
half a century.

In regard to the medicinal virtues of the
Pine-apple, it is counted very nouriſhing,
to obtund acrimony, and thereby allay tickling coughs ; but Tournefort ſays, that too
liberal an uſe of them has often been attended with bad conſequences, by putting the
blood into a violent fermentation ; and indeed this is the caſe with almoſt all the tropical fruits.

5 Bromelia karatas. *The Penguin.*
Lin. Sp. pl. 408.

This is a perennial plant, and a native of
the Spaniſh Weſt-Indies. It ſends forth a
multitude of hard, ſtiff leaves, ſtanding
cloſe to the root, and when fully grown are
eight or nine feet high, two or three inches
broad, and ſtudded with ſharp, hooked ſpines
on their margins. The edges roll inward,
in the manner of ſome of the Aloes, by
which means they ſerve as ſo many gutters
to convey the rains and dews to the root.

In

In the centre of this large tuffock of leaves, and near the ground, there grows a circular crown, of about a foot diameter, from which comes a clufter of fruit, each when feparated much the fize of ones finger, but are pointed at both ends, and are quadrangular in the middle, whereby they are fo neatly fitted to each other, that they cannot eafily be parted, unlefs thoroughly ripe. They are clothed with a fmooth, and almoft cream-coloured hufk. Within this hufk is contained a white pulpy fubftance, which is the edible part, and if the fruit be not perfectly ripe, it has fome fmall flavour of the *Pine-apple*. The juice is very auftere in the ripe fruit, and is made ufe of to acidulate punch. The inhabitants in the Weft-Indies make a wine from this fruit, which is very intoxicating, and has a good flavour, but it will not keep long before it runs into a ftate of putrefaction.

The phyfical virtues of the *Penguin* are to cool and quench thirft, and a moderate ufe of them has been found highly ferviceable in fevers.

6 Cactus opuntia. *Prickly Pear. Lin. Sp. pl.* 669.

Ficus indica, folio fpinofo, fructu majore. *Bauh. Pin.* 458.

This perennial is a native of Peru and Virginia. It here goes by the name of

N 3 *Common*

Common Indian Fig. The plant in its natural state rises with a thick, strong stem,
but being propagated here by setting its
leaves in the ground, the whole plant with
us is only a series of these leaves, or rather
branches, shooting out of the sides and ends
of each other. These are of an oval form,
compressed, and somewhat resemble flatted,
green Figs. The flowers come out at the
extremities of the leaves or branches, sitting
upon the embryo of the fruit, and are composed of several concave petals that spread
open in a double row; they are of a pale
yellow colour, and include many stamina,
tipped with oblong summits, and one style
crowned with a pointed stigma. When the
flower falls, the embryo swells to an oblong
fruit, about the size of a middling Plumb, of
a red purple colour within, of a pale yellow
without, is set with small spines in clusters,
and contains many small roundish seeds.

These fruits are very pleasant to the palate,
and of a cooling nature. Mr. Dampier,
who experienced it upon the spot where the
plants grew naturally, says, that by eating
a few of them the urine will be tinctured
as red as blood. It has been generally supposed that this is the plant upon which the
insect, called *Cochineal,* feeds; but this is a
mistake, for that little creature lives on the
Cactus cochinillifer, so named after the animal.

7 CACTUS

7 CACTUS triangularis. *True Prickly Pear. Lin. Sp. pl.* 669.

Cactus debilis brachiatus æqualis triquetrus fcandens five repens, fpinis breviffimis confertis, *Browne's Jam.* 468.

This grows both in Brazil and Jamaica, and is there planted near their houfes for the fake of its fruit. It hath weak, triangular, creeping ftalks, which ftrike root at their joints, and by which they may be trained up to a great height. Thefe divide into many equal branches, almoft covered with very fhort fpines in clufters. The flower is compofed of a multitude of narrow, fharp-pointed petals, which fpread open like thofe of the Sunflower, and when fully expanded, form a circle of nine or ten inches diameter; but they are of fhort duration, not lafting more than five or fix hours.

The fruit is round, red on the outfide, about the fize of a Bergamot Pear, of a moft delicious flavour, and in great efteem among the inhabitants.

8 CAPSICUM annuum. *Annual Guinea Pepper. Lin. Sp. pl.* 270.

Piper indicum vulgatiffimum. *Bauh. Pin.* 162.

The *Annual Guinea Pepper* is a native of America, but on account of the beautiful colour of its pods, or more properly berries, it is now cultivated in almoft every garden

N 4   in

in England. It varies prodigiously in regard
to the fize, form, and colour of its fruit;
fome being very long, bent and fharp
pointed; others are fhort, obtufe, or heart-
fhaped, and of other forms. In refpect to
colour, fome are of a fine fcarlet, fome of
an orange, and others of a light yellow.
This plant is cultivated greatly in the Carib-
bee Iflands, where the inhabitants, and alfo
the Negroes, ufe the pods in almoft all their
foups and fauces, and by reafon the flaves
are exceedingly fond of them, the whole
genus has acquired the name of *Guinea
Pepper.*

These pods or berries make an excellent
pickle, and there is one variety which Miller fays is preferable to the reft for this purpofe. His words are: " The pods of this
fort are from one inch and an half, to two
inches long, are very large, fwelling, and
wrinkled; flatted at the top, where they
are angular, and fometimes ftand erect, at
others grow downward. When the fruits of
this fort are defigned for pickling, they
fhould be gathered before they arrive to their
full fize, while their rind is tender; then
they muft be flit down on one fide to get out
the feeds, after which they fhould be foak-
ed in water and falt for two or three days;
when they are taken out of this and drained,
boiling-vinegar muft be poured on them, in
a fufficient quantity to cover them, and
clofely

clofely flopped down for two months; then
they fhould be boiled in the vinegar to make
them green; but they want no addition of
any fort of fpice, and are the moft wholefome and beft pickle in the world." This
fort Miller calls *Bell-pepper*.

9 CAPSICUM frutefcens. *Perennial Guinea Pepper*. *Lin. Sp. pl.* 271.

Piper filiquofum magnitudinis baccarum
afparagi. *Bauh. Hift.* 2. *p.* 944.

This is a fhrubby plant, and rifes four or
five feet high, breaking into many branches,
furnifhed with narrow, lance-fhaped leaves.
Like the foregoing, it varies in the form and
colour of its fruit; they being oval, roundifh, or pyramidal in different plants, and of
a yellow or a red colour. Their fize is
nearly that of a Barberry. It is a native of
the Eaft-Indies, but is much cultivated in
the Weft, where they have a variety of it
with an oval, red fruit, which they call
Bird-pepper; the berries of this variety they
pickle, but the principal ufe they put them
to, is to make the famous *Cayan Butter*,
called alfo *Pepper-pot*. In order to this
they dry the berries, beat them to a powder;
and mixing fome other ingredients among
them, the whole is kept and ufed occafionally in their fauces, and is efteemed the beft
of all fpices. Thefe *Pepper-pots* are often
fent

fent to England and other places, and generally meet with an equal approbation.

10 CARICA papaya.. *The Papaw.* *Lin.*
Sp. pl. 1466.

Carica fronde comofa, foliis peltatis; lobis varié finuatis. *Browne's Jam.* 360.

This tree is a native of both the Indies,
alfo of the Gold-coaft of Africa, and is male
and female in diftinct plants. It fends up
a hollow, herbaceous ftem, to the height
of fifteen or eighteen feet, and about feven
inches in diameter. Near the top the leaves
come out on all fides the ftem, and are fupported on long foot-ftalks; they are divided into feveral lobes, which are again cut
into many irregular fegments. The flowers
are produced in loofe bunches from the bofoms of the leaves; thofe of the male are
white, funnel-fhaped, cut at their brims
into five parts, and have ten ftamina each,
five of which are alternately fhorter than
the reft. The female flowers are yellowifh,
and compofed of five long, narrow petals,
including a very fhort ftyle, crowned by
five oblong ftigmata. Thefe are fucceeded
by fruit of different fhapes and fizes; fome
being angular, and about as big as middling
Pears; others are compreffed at both ends,
and about the fize of a fmall Squafh; whilft
fome are globular, oval, or conical. They
contain

contain numerous feeds, which are egg-
fhapped and furrowed. The fruit, and all
the other parts of the tree abound with a
milky, acrid juice, which is applied for
killing of ringworms.

When the roundifh fruit are nearly ripe,
the inhabitants of India boil and eat them
with their meat, as we do Turneps. They
have fomewhat the flavour of a Pompion.
Previous to boiling they foak them for fome
time in falt and water, to extract the corrofive juice; unlefs the meat they are to be
boiled with fhould be very falt and old, and
then this juice being in them will make it
as tender as a chicken. But they moftly
pickle the long fruit, and thus they make
no bad fuccedaneum for mango. The buds
of the female flowers are gathered, and made
into a Sweet-meat; and the inhabitants are
fuch good hufbands of the produce of this
tree, that they boil the fhells of the ripe
fruit into a repaft, and the infides are eaten
with fugar in the manner of Melons.

The ftem being hollow, has given birth
to a proverb in the Weft-India Iflands;
where, in fpeaking of a diffembling perfon,
they fay he is as hollow as a *Popo*.

11 CARICA pofopofa. *Pear-fhaped Papaw. Lin. Sp. pl.* 1466.
Carica fylveftris minor, lobis minus divi-
fis,

fis, caule fpinis inermibus. *Browne's Jam.*
360.

This is a fhrubby tree, and a native of
Surinam, in South America. The ftem
breaks into feveral branches, furnifhed with
leaves fomewhat like thofe of the former,
but the lobes are fmaller, and not finuated.
The flowers are of a rofe colour, and are
fucceeded by Pear-fhaped fruit, of various
fizes, fome being near eight inches long,
and three thick, and others not above half
as large. They are yellow both without
and within, and of a fweeter flavour than
the common Papaw.

12 CHRYSOPHYLLUM cainito. *Star Ap-*
ple. Lin. Sp. pl. 278.

Cainito folio fubtus aureo, fructu mali-
formi. *Plum. gen.* 10.

This is a native of the warm parts of
America, and grows to the height of thirty
or forty feet, dividing towards the top into
many flender, pendulous branches, fet with
entire, oblong-oval, ftriated leaves, covered
with a ruffet-coloured down underneath,
and ftanding alternately, on footftalks.
Thefe, when the fun fhines, glifter like a
gold-coloured fattin. The flowers are produced at the extremities of the branches, in
large bunches; and each is compofed of a
fmall quinquefid calyx, and a bell-fhaped
petal, cut into five fegments at their brims,
including

including five awl-fhaped ftamina, tipped
with twin fummits, together with one ftyle,
crowned with a quinquefid ftigma. The
germen is roundifh, and grows to the fize
of a fmall Apple. The fruit is fmooth, of
a purple colour, and contains four or five
black, roughifh feeds. There is a variety
of this tree with fruit the fhape of an olive.

Thefe apples, when frefh off the tree,
have an auftere, aftringent tafte; but if laid
up fome time to mellow they acquire an
agreeable flavour, and are much efteemed.

13 CHRYSOPHYLLUM glabrum. *Sapadillo*. *Lin. Sp. pl.* 278.

This too is a native of America, but is a
much fmaller tree than the former; the
leaves are very fmooth on both fides, the
flowers are produced at the fides of the
branches, and the fruit is about the fize of a
Bergamot Pear. This contains a white
clammy juice, when frefh, but after being
kept a few days, it becomes fweet, foft and
delicious. Inclofed are four or five black
feeds, about the fize of thofe of a Pomkin.

14 CITRUS medica. *Common Citron. Lin.
Sp. pl.* 1100.

Malus medica. *Bauh. Pin.* 435.

The *Common Citron* grows naturally in
many parts of Afia. The leaves are broad
and ftiff, like thofe of the Laurel, and with-

out

out an appendage to the footstalks, it being linear. The flower hath a monophyllous calyx, cut into five teeth, and five oblong petals, which expand in the form of a Rose. It hath ten unequal stamina, joined in three bodies at their base, and a cylindrical style, crowned with a round stigma. The germen is oval, and becomes an oblong fruit, with a thick fleshy rind, and having many cells, containing two oval feeds each. Linnæus makes the *Lemon* and the *Lime-tree* only varieties of this, but both these have generally twelve or more stamina in their flowers, joined in three or four bodies. The varieties now raised by sowing the feeds of these three forts are almost numberless. They are all excellent fruits, very grateful to the stomach, and proper for allaying drought in fevers. The *Florentine Citron*, (which is a sharp-pointed fruit, bent at the ends, and covered with a warted rind) Miller fays, is of such odoriferous smell, and fine flavour, that a single fruit commonly fold at Florence for two shillings.

15 CITRUS aurantium. *The Orange.* *Lin. Sp. pl.* 1100.

Malus aurantia major. *Bauh. Pin.* 436.

The *Orange-tree* is a native of the East-Indies. The chief specific differences between this and the *Citron* are; the footstalk of the leaf of the *Orange* is winged at its

base,

bafe, or has an heart-fhaped appendage, whereas that of the *Citron* has none, but is all the way of a breadth ; the flower of the *Orange* has many more ftamina than that of the *Citron*. Thefe trees are ever-greens, and in their native foils have bloffoms and fruit the year round. There are many varieties cultivated of the *Orange*; but as they cannot be produced here to perfection, without much expence, I fhall forbear fetting them down, and only obferve that the fmall *Curaffao Oranges*, fold in the fhops, are the young fruit of the Seville Orange dried.

16 CITRUS ducumanus. *Shaddock Orange.* *Lin. Syft. Nat.* 508.

Malus aurantia fructu rotundo maximo pallefcente caput humanum excedente. *Sloane's Jam.* 212. *Hift.* I. *p.* 41.

Linnæus formerly made this only a variety of the *aurantium*, the largenefs of the fruit not being a fufficient mark with him to conftitute a fpecific difference ; but it has been found that both the leaves and flowers are larger, and that the latter are produced in a racemus, which is a little downy. This plant was brought from the Eaft Indies to the Weft, where it is now much cultivated, and fometimes produces fruit larger than a man's head, but they are of an harfh flavour, and pale colour, when compared with thofe

of

of India, the flesh of which is sweet, and of a deep gold colour.

17 CRATEVA marmelos. *Bengal Quince. Lin. Sp. pl.* 637.

Cydonia exotica. *Bauh. Pin.* 435.

This is a large tree, and grows spontaneously in several parts of India. It breaks into many branches, armed with long, sharp spines in pairs, and are furnished with trifoliate, oblong leaves, ending in an acute point. The flowers are produced from the sides of the branches, in small clusters of six or seven together, upon a common footstalk, each flower consisting of five acute, reflexed petals, of a green colour on their outside, but white within, surrounding many stamina, which are longer than the petals, and one long, incurved style. The germen is oval, and swells to a roundish fruit, including many kidney-shaped seeds.

The fruit is about the size of an Orange, and covered with a hard bony shell, containing a yellow, viscous pulp, of a most agreeable flavour ; this is scooped out, and being mixed with sugar and orange, is brought to the tables of the grandees in India, who eat it as a great delicacy, and also esteem it as a sovereign remedy against dysenteries.

18 DIOSPYROS

18 Diospyros lotus. *Indian Date Plum.*
Lin. Sp. pl. 1510.

Lotus africana latifolia. *Bauh. Pin.* 447.

This tree grows in Italy, and fome other places in the fouth of Europe, but is fuppofed to have been originally brought thither from Africa. It rifes to a confiderable height, dividing towards the top into many branches, which are furnifhed with oval, fharp-pointed leaves, beautifully variegated on their upper furface. Some trees bear all hermaphrodite flowers, and others produce only male. The hermaphrodite flowers have a lafting calyx, divided into four parts, including a pitcher-fhaped petal, with eight ftamina, joined to the calyx, and a roundifh germen in the centre, fupporting a long ftyle, crowned with an obtufe, bifid ftigma. The flowers come out in a fcattered order upon the branches, and are fucceeded by large globular berries, divided into eight cells, each including one long, compreffed feed. The male flowers are formed like the others, but want the germen. There is a variety of this tree with narrow leaves.

Thefe Plums are grateful to the palate; they are by many fuppofed to be the fame fort of fruit as thofe which tempted the companions of *Ulyffes*, and with which they were fo infatuated, that it was with difficulty they were forced from the trees to their fhips.

O   19 Dios-

19 DIOSPYROS virginiana. *Pifhamin Plum.*
Lin. Sp. pl. 1510.

Loti Africanæ fimilis indica. *Bauh. Pin.*
448.

The trivial name of this fpeaks it to be a
native of Virginia. It is a fmaller tree
than the former, feldom rifing more than
fourteen feet, whereas the *lotus* often gets
to thirty. This divides near the ground
into irregular branches, furnifhed with long,
narrow leaves, of the fame colour on both
fides.

The fruit of this fpecies are not eatable
frefh off the tree, but like Medlars mult be
kept fome time, and then they have a good
flavour.

20 FICUS carica. *Common Fig. Lin.*
Sp. pl. 1513.

Ficus communis. *Bauh. Pin.* 457.

The *Fig-tree* is a native of Afia, but is
now cultivated almoft all over Europe,
whereby it is fo well known as to need no
defcription. The fructification of the *Fig* is
exceedingly curious, and deferves particular
notice, for here the parts of generation are
contained within the berry, which thereby
becomes both a *pericarpium,* and a covered
receptacle of flowers. The fruit of the
Wild Fig, called *Caprificus,* contains both
male and female flowers, on diftinct peduncles. The male flowers, which are but few

in

in number, are placed in the upper part of the fruit, each having a trifid calyx, containing three briftly ftamina. The female flowers are very numerous, ftand upon feparate peduncles below the males, and each confifts of a quadrifid calyx, having one ftyle. Thefe wild fruits are not eatable, for they never perfectly ripen, but are faid to be abfolutely neceffary for ripening the garden *Fig*, or rather to fecundate it, and prevent its falling off; for the cultivated *Fig* is moftly found to contain female flowers only. The manner of effecting this fecundation, as related by naturalifts, and which is called *Caprification*, is briefly as follows:

In the Greek iflands, where they cultivate *Figs* for a crop, there grow many *Wild Fig-trees*, in the fruit of which breed fmall infects of the gnat kind. Thefe little creatures, in their worm ftate, feed upon the kernels of the fig-feeds, and are nourifhed in the fruit till they are transformed into flies, when piercing the coats of the *Figs*, they iffue forth, copulate, repair to other *Fig-trees*, which are then in flower, and pricking the fruit, enter by the apertures they make, range among the flowers in the infide, and depofite their eggs. Now it is fuppofed that thefe gnats bring with them about their bodies the fertilizing duft of the male flowers of the *Wild Figs*, and after

O 2   they

they get an entrance, they scatter it upon the germina of the female flowers of the cultivated ones, and thereby impregnate the seeds, which causes the fruit to stand, and ripen much better and sooner. These effects having been seen to happen upon the intercourse of the gnats with the different trees, put the husbandmen upon a method of rendering them subservient to their own purposes, and *Caprification* is become a main article in the cultivation of *Figs*; for, that the growers may make sure of their crops, they collect these insects, and place them upon the branches of their trees; or they cut off the Figs of the wild trees and hang them about their domestic ones, the fruit of which the gnats readily enter, and, as before observed, sprinkle the dust they bring with them upon the female flowers in the inside of these fruits, by which means they become fecundated.

The varieties of the Fig are very numerous, but several of them are not worth cultivating. Those most deserving attention in England are the following :

1 The *Brown Ischia.*
2 The *Black Genoa.*
3 The *Small White.*
4 The *Large White Genoa.*
5 The *Black Ischia.*
6 The *Malta.*
7 The *Brown Naples, or Murrey.*
8 The *Green Ischia.*
9 The *Brunswick.*
10 The *Long Brown Naples.*

The

The *Brown Ifchia* is a very large Fig, of a globular form, has a large eye, and is pinched in near the footftalk. It is of a chefnut-brown colour on the outfide, purple within, hath large grains, and a fweet, high-flavoured-pulp. It ripens early in Auguft, and is fubject to burft.

The *Black Genoa* is a longifh Fig, with a fwelled obtufe top, but is very flender towards the ftalk. It is of a black purple colour on the outfide, covered with a purple flue; the infide is of a bright red, and the pulp hath a high flavour. Ripe early in Auguft.

The *Small White* is a roundifh Fig, with a very fhort footftalk, and is flattifh at the crown. The fkin is thin, and of a pale yellow colour when ripe. It is white in the infide, and the flefh is very fweet. In perfection in Auguft.

The *Large White Genoa* is a roundifh Fig, a little lengthened toward the ftalk. This too is yellowifh when ripe, but it is red within. Ripe with the former.

The *Black Ifchia* is a middling fized Fig, rather fhort, and a little flatted at the crown. It is black on the outfide, and of a deep red within. The pulp has a rich flavour. It ripens in Auguft.

The *Malta* is a fmall brown Fig, much flatted at the crown, and greatly pinched in toward the ftalk. It is brownifh both outfide and in. The pulp or flefh is fweet and well flavoured. Ripe with the former.

The *Brown Naples* is a pretty large round Fig, of a light brown colour on the outfide, with a few marks of a dirty white. The infide is nearly of the fame colour, the feeds are large, and the flefh is well flavoured. Ripe toward the end of Auguft.

The *Green Ifchia* is an oblong Fig, but is roundifh at the crown. The outfide is green, but when

fully

fully ripe has a brownifh caft. The flefh is purple, and well flavoured. It ripens with the former.

The *Brunfwick* is a pear-fhaped Fig, of a large fize, of a brown colour on the outfide, and of a lighter brown within. The flefh is coarfe, and not highly flavoured. Ripe at the beginning of September.

The *Long Brown Naples* hath a long footftalk, and the Fig is a little flatted at the crown. When ripe the fkin is of a dark brown colour, the feeds are large, the flefh inclining to red, and is well flavoured. Ripe in September.

In the iflands of the Archipelago, where *Caprification* is univerfally practifed, they dry their Figs in ovens, to kill the infects and their eggs; this much hurts the flavour of the fruit, but neverthelefs they are the chief fupport of the peafants and monks there, in conjunction with Barley-bread. With refpect to the virtues of Figs, they are faid to inflame the blood, if eaten frefh off the trees; but dried, they are of an emollient nature, and good in diftempers of the breaft, and defluxions of rheum upon the lungs.

21 FICUS fycomorus, *Pharaoh's Fig. Lin. Sp. pl.* 1513.

Ficus folio mori, fructum in caudice ferens. *Bauh. Pin.* 459.

This is a native of Egypt. It is a large tree, dividing into many fpreading branches, plentifully furnifhed with leaves, fhaped

like

like thofe of the Mulberry. The fruit are
not produced on the fmall fhoots, but from
the trunk and thick branches. They are
fhaped like thofe of the common Fig, but
are far inferior in flavour, and not much
efteemed.

The wood of this tree is but of a fpungy
nature, yet the ancient Egyptians made ufe
of it for coffins to contain their mummies,
fome of which are ftill to be found in their
catacombs, or fubterraneous burying places,
where they are placed upright, and have
been depofited more than three thoufand
years.

22 GARCINIA mangoftana. *The Mangofteen. Lin. Sp. pl.* 635.

Laurifolia javanenfis. *Bauh. Pin.* 461.

This tree is a native of the ifland of Java,
and is alfo found in the Molucca Iflands.
It fends up a ftraight, tapering ftem, to
eighteen or twenty feet, having branches
coming out on all fides from near the bottom, and continuing to diminifh equally in
length to the top, whereby they form the
tree into a compleat cone. The leaves are
long, pointed at both ends, fmooth, of a
lucid green on their upper fide, and of an
olive colour on their back. The flower is
compofed of four almoft round petals, nearly
refembling the Rofe in colour. The calyx
is of one piece, and on expanding breaks
into four lobes. In the centre of the flower

O 4   is

is one very fhort ftyle, crowned with an
octifid ftigma, and furrounded by fixteen
erect ftamina, having globular fummits.
The germen is roundifh, and becomes a
berry of the fize of an Orange, covered with
a thick rind, of a brown purple, mixed
with a greyifh green on the outfide, but of
a rofe colour within, and contains eight
hairy, flefhy, angular feeds.

According to the concurring teftimonies
of all travellers, this fruit is the moft excellent flavoured, and the moft falubrious,
of any yet known; it being fuch a happy
mixture of the tart and the fweet. *Rum-
phius* fays, the flefh is juicy, white, almoft
tranfparent, and of as delicate and agreeable a flavour as the richeft Grapes. Both
tafte and fmell is fo grateful, that it is almoft impoffible to be cloyed with eating it;
and that when fick people have no relifh
for any other food, they generally eat this
with great delight; but fhould they refufe
it, their recovery is no longer expected. It
is remarkable too, fays he, that this fruit is
eaten with fafety in almoft every diforder.
The bark, he adds, is ufed with fuccefs in
the dyfentery and tenefmus; and an infufion
of it is efteemed a good gargle for fore
mouths, or ulcers in the throat. The Chinefe dyers ufe this bark for the bafis of a
black colour, in order to fix it the firmer.

23 MORUS nigra. *Lin. Sp. pl.* 1398.

Morus

Morus fructu nigro.   *Bauh. Pin.* 459.

The *Black Mulberry* grows naturally on the coast of Italy.   The tree is well known by being frequent in our gardens, nor need any thing be obferved in regard to the excellent flavour of its fruit.   Thefe furnifh the fhops with a fyrup, which is of a cooling, aftringent nature, and is much ufed in gargarifms for fore mouths.

24 MORUS rubra.   *Lin. Sp. pl.* 1399.

The *Red Mulberry* is a native of Virginia. It differs from the common *Black Mulberry* in the leaves being longer and rougher, and in the catkins being cylindrical.   When the leaves firft expand they are very hairy underneath, fometimes palmated, but oftener trilobed and a little hairy.   The catkins are about the length of thofe of the Birch-tree.

25 MORUS alba. *White Mulberry.* *Lin. Sp. pl.* 1398.

Morus fructu albo.   *Bauh. Pin.* 459.

This differs from the others not only in the fruit being white, but its leaves are obliquely heart-fhaped, and fmooth.   It is a native of China, where it is cultivated more for its leaves than its fruit, for the purpofe of feeding Silkworms; but though this is the practice in China, yet it has been proved by experiments, that the leaves of the *nigra* are far preferable for this ufe, and

that

that the worms which had been fed with the latter, always produced much better filk, than thofe which were fed with the former. Thefe creatures are more fond of the leaves of the black than of the white *Mulberry*, and if they be kept any time on the white, and then put to the black, they will feed till they burft.

26 Musa paradifiaca. *Plantain - tree.*
Lin. Sp. pl. 1477.

Ficus indica, fructu racemofo, folio oblongo. *Bauh. Pin.* 508.

The *Plantain-tree* grows fpontaneoufly in many parts of India, but has been imme-morially cultivated by the Indians in every part of the continent of South America. It is an herbaceous tree, growing to the height of fifteen or twenty feet. Its ftem, which is about eight inches thick at the bottom, and regularly tapers to the top, is enwrapped with many leafy circles; thefe expand at the extremity of the trunk, and form the footftalks, and midribs of the leaves, which come out on every fide. The leaves are fmooth, of an oval form, in co-lour like thofe of Cabbage, five or fix feet long, and two broad, have many tranfverfe, prominent veins, but the leafy part is fo thin, that a ftrong wind often tears them into rags, and makes them cut an uncouth figure. On the firft appearance of the leaves
they

they are rolled up like the young fhoots of a
Brake; but as they advance, they turn
backward, and their growth is fo quick,
that it may be almoft feen by a perfon nigh.
From among the leaves comes forth a long
fpike of flowers, in circular bunches; thofe
at the upper part of the fpike are all male,
and thofe at the bottom all hermaphrodite.
Each of thefe bunches has its fpatha, of an
oblong-oval form, and a fine purple colour.
The flowers are of the lip kind, the petals
forming the upper, and the nectarium * the
under lip. Each flower has fix ftamina,
five feated in the petals, and the fixth in
the nectarium. The germen is placed below the flower; it is very long, nearly triangular, fupports a cylindrical ftyle, longer
than the petals, and is crowned by a roundifh
ftigma.

The fruit are nearly of the fize and fhape
of ordinary Cucumbers, and when ripe of a
pale yellow colour, of a mealy fubftance, a
little clammy, a fweetifh tafte, and will dif-
folve in the mouth without chewing. The
whole fpike of fruit often weighs forty or
fifty pounds. When they are brought to
table by way of defert, they are either raw,
fried, or roafted; but if intended for bread,
they are cut before they are ripe, and are
then either roafted or boiled. The trees

---

* The Nectarium is a gland, or appendage to the petal,
and is appropriated for containing the honey.

being

being tall and flender, the Indians cut them down to get at the fruit; and in doing this they fuffer no lofs, for the ftems are only one year's growth, and would die if not cut; but the roots continue, and new ftems foon fpring up, which in a year, produce ripe fruit alfo. From the ripe *Plantains* they make a liquor, called *Miflaw*; when they make this, they roaft the fruit in their hufks, and after having totally beat them to a mafh, they pour water upon them, and as the liquor is wanted, it is drawn off. But the nature of this fruit is fuch, that they will not keep long without running into a ftate of putrefaction, and therefore in order to reap the advantage of them at all times, they make cakes of the pulp, and dry them over a flow fire; and as they ftand in need of *Miflaw*, they mafh the cakes in water, and they anfwer all the purpofes of the frefh fruit. Thefe cakes are exceedingly convenient to make this liquor of in their journies, and they never fail to carry them for that purpofe. The leaves of the tree being large and fpacious, ferve the Indians for table-cloths and napkins.

27 Musa fapientum. *The Banana. Lin. Sp. pl.* 1477.

Mufæ affinis altera. *Bauh. Pin.* 580.

This is a native of both the Indies, and is much cultivated in the American iflands,

by

by the name of *Banana*. It differs from the former in the stem being marked with purple stripes, in the other not; in the fruit being shorter, straighter, and more obtuse. These grow in bunches from ten to fourteen pounds. They have a fragrant smell, and an agreeable delicious taste, far preferable to the *Plantain*, but yet inferior to many European fruits.

The leaves of this tree are by many authors supposed to be the same sort with those our first parents made themselves aprons. They indeed are called in Scripture Fig leaves; but as these are larger and more fit for the purpose than any species of Fig, there is the greater probability in the supposition; these being four or five feet long, one broad, and of a pretty tough texture.

28 MESPILUS germanica. *The Medlar.*
Lin. Sp. pl. 684.

Mespilus germanica, folio laurino non serrato. *Bauh. Pin.* 453.

This grows naturally in Sicily, but is so common in gardens, and orchards, as to make it generally known. Linnæus makes the Dutch Medlar only a variety of this, though many think it a distinct species. The Dutch is the sort now chiefly cultivated, by reason it produces larger and better

flavoured

flavoured fruit; but neither of them are eatable, unlefs kept till they be rotten.

29 MAMMEA Americana. *The Mammee.* *Lin. Sp. pl.* 731.
Arbor indica Mamei dicta. *Bauh. Pin.* 417.

This grows naturally in Jamaica, and in many parts of the Spanifh Weft Indies. It rifes to near feventy feet, with a ftraight ftem, deftitute of knots and branches, except at the top, where it breaks into rough boughs, furnifhed with oblong, obtufe, fhining green leaves, which continue the year through. The flowers are compofed of four concave, fpreading petals each, furrounding many fhort, hair-like ftamina, having oblong fummits, and one cylindrical ftyle, crowned with a convex ftigma. The germen is roundifh, and becomes a globular, yellowifh, rough fruit, about the fize of a Quince, and contains three or four almoft oval feeds, about as big as almonds.

Thefe fruits have a very grateful flavour, and are much cultivated in Jamaica, where they are generally fold in the markets for one of the beft the ifland produces.

30 MALPHIGIA glabra. *Barbadoes Cherry. Lin. Sp. pl.* 609.
Malphigia fruticofa erecta, foliis nitidis
5   ovatis

ovatis acuminatis, floribus umbellatis, ramulis gracilibus. *Browne's Jam.* 230.

This grows naturally in Jamaica, Brafil,
Surinam, and Curaçoa, but it is now cultivated in moft of the Weft-India Iflands,
and particularly at Barbadoes. It fends up
a flender trunk to about fifteen feet covered
with a light brown bark. At the top it
breaks into many branches, the twigs of
which are furnifhed with oval, fmooth,
acute pointed leaves in pairs. The flowers
are produced in bunches, upon long peduncles; they confift of five kidney-fhaped,
rofe-colour petals each, joined at their bafe,
and include ten awl-fhaped, erect ftamina,
tipped with heart-fhaped fummits. The
germen is fmall and roundifh, and fupports
three flender ftyles, crowned with obtufe
ftigmata.

The berries are red, about the fize of
fmall Cherries, and are gathered and eaten
by the inhabitants, the fame as Cherries
are in England, but they are far inferior.

31 MALPHIGIA punicifolia. *Pomegranate-
leaved Malphigia. Lin. Sp. pl.* 609.

Malphigia fruticofa erecta, ramulis gracilibus patentibus, floribus folitariis. *Browne's
Jam.* 230.

This is a native of Jamaica. It is a fmaller
tree than the former, and grows after the
manner of a fhrub. The branches are
flender,

flender, fpreading, covered with a light brown bark, and are furnifhed with leaves like thofe of the Pomegranate. The flowers are produced fingle in this fpecies, contrary to thofe of the firft, which come out in umbels. The fruit are rather more acid than the former, but are eaten after the fame manner.

32 PASSIFLORA maliformis. *Apple-fhaped Granadilla. Lin. Sp. pl.* 1355.

Paffiflora foliis cordato-oblongis integerrimis, floribus folitariis, involucro tripartito integerrimo. *Roy. lugdb.* 261.

This is a native of the Ifland of Dominica in the Weft-Indies, and is cultivated both for ornament and ufe in feveral of the Iflands there. It fends forth an herbaceous, climbing ftalk, having tendrils at every joint, by which it faftens to the hedges for fupport, and runs to the length of near twenty feet. There is alfo at each joint one oblong heart-fhaped leaf, having two glands upon its footftalk. The flowers are produced fingly at the footftalks of the leaves, upon long peduncles, and each is compofed of a three-leafed, red cover, enclofing five white petals and numerous blue rays, which fpread very wide, and make a moft beautiful appearance; but they are of fhort duration. When the flower falls, the germen fwells to a yellow berry, of the fize and fhape of a

3   fmail

ſmall Apple, containing a ſweet pulp, and many oblong, browniſh ſeeds.

Theſe berries have a pleaſant flavour, and are generally ſerved up at table by way of deſert.

33 PASSIFLORA laurifolia. *Bay-leaved Paſſion-flower. Lin. Sp. pl.* 1356.

Paſſiflora foliis ſolitariis oblongis integerrimis, floribus ſolitariis, involucro tripartito dentato. *Roy. lugb.* 532.

The *laurifolia* is a native of Surinam, the fruit of which is greatly beneficial to the inhabitants of that hot climate. It ſends forth many tough, ſlender ſtalks, with claſpers at their joints, by which they climb up the trees and buſhes to the height of twenty feet or more. Theſe are furniſhed with oblong-oval, entire leaves, reſembling thoſe of the Laurel, and having two glands on their footſtalks. The flowers are produced at the joints of the ſtalks, in manner of the former. Their full-grown buds are nearly as large as thoſe of the garden ſingle Poppy, each having a cover, compoſed of three indented oval, green leaves; theſe encloſe the flower-cup, which conſiſts of five pale green, oblong leaves, with white inſides. The petals are white, ſpotted with brown, and are but little more than half the breadth of the leaves of the calyx or cup. The rays of the flowers are of a violet co-

P   lour,

lour, the column in the centre is yellowiſh, its germen at the top the ſame, but the three ſtyles are purple. On the fading of the flower, the germen ſwells to a yellow, oval berry, ſomewhat reſembling a Citron, but ſmoother.

The fruit of this ſpecies have a delicate acid flavour, far preferable to the former, and are excellent for quenching thirſt, abating heat in the ſtomach, encreaſing the appetite, recruiting the ſpirits, and allaying the heat in burning fevers.

34 PSIDIUM pyriferum. *Pear Guava.* *Lin. Sp. pl.* 672.

This grows naturally in both the Indies, and is much cultivated in the American Iſlands. It riſes to eighteen or twenty feet, dividing into many branches from near the bottom; theſe are covered with a reddiſh-gray bark, are angular, and furniſhed with narrow, bluntiſh leaves, three or four inches long, ſupported on ſhort footſtalks: from the wings of theſe the flowers are produced ſingly on peduncles, about an inch long; each is compoſed of five white, concave petals, inſerted in a bell-ſhaped calyx, cut at the brim into four or five teeth, and of numerous ſhort ſtamina, tipped with ſmall, pale yellow ſummits. The germen is roundiſh, ſeated below the calyx, and ſupports a very long awl-ſhaped ſtyle, crowned with a

2   ſimple

simple ftigma; it grows to a whitifh Pear-
fhaped berry, adorned at the apex with the
remains of the calyx, and includes many
fmall feeds.

35 PsIDIUM pomiferum.   *Apple Guava.*
Lin. Sp. pl. 672.

Guajabo pomifera indica, pomis rotundis.
Bauh. Pin. 437.

This and the former are promifcuoufly
defcribed by travellers as one fpecies only,
but Linnæus has plainly pointed out two
diftinct ones.   The leaves of the *pomiferum*
are fharp-pointed, in the *pyriferum* they are
rather obtufe; the latter has only one flower on a peduncle, but the former has three.
The fruit of the *pyriferum* is fhaped like a
Pear; but that of the *pomiferum* like an
Apple.   This laft is the fort moft cultivated, the pulp having a fine acid flavour,
whereas the former is fweet, and therefore
not fo agreeable in hot climates.

Of the inner pulp of either fort the inhabitants make jellies; and of the outer
rinds they make tarts, marmalades, &c.
The latter too they ftew, and eat with
milk, and prefer them to any other ftewed
fruits.   They have an aftringent quality,
which fhould forbid ftrangers making too
free with them, as they are apt to render
the body coftive.   This aftringency runs
through all parts of the trees, and exifts

pretty

pretty copioufly in the leaf-buds, which are occafionally boiled with barley and liquorice, as an excellent drink againft diarrhæas. A fimple decoction of the leaves, ufed as a bath, are faid to cure the itch, and moft cutaneous eruptions.

· 36 SOLANUM lycoperficum. *Lin. Sp. pl.* 265.

Solanum pomiferum, fructu rotundo ftriato molli. *Baub. Pin.* 167.

The *Love Apple* is an annual and a native of America. It hath herbaceous, branching, trailing, hairy ftems, four or five feet long, and without fpines. Thefe are furnifhed with pinnated leaves, of an offenfive fmell, and each is compofed of four or five pair of jagged pinnæ, ending in an acute point. The flowers come out in long racemi in different parts of the branches; they are yellow, monopetalous *, plaited, cut at their brims into five fharp teeth, and have five fmall awl-fhaped ftamina, clofely furrounding a flender ftyle, which fits upon an oval germen. As the flower withers, the germen fwells to a round, fmooth berry, bigger than a large Cherry, and of various colours on different plants; on fome it being red, on others of a deep orange, and on fome yellow.

* Confifting of one petal.

2

That

That which is fo much cultivated by the Portugueze Linnæus makes only a variety of this. They call it *Tomatas*, and it differs from the original in the fruit being deeply furrowed. Thefe berries are in fuch efteem both among the Portugueze and the Spaniards, that they are an ingredient in almoft all their foups and fauces, and are deemed cooling and nutritive.

37 SOLANUM melongena. *Lin. Sp. pl.* 266.

Solanum pomiferum, fructu oblongo. *Bauh. Pin.* 167.

- The *Mad Apple* is a native of Afia, Africa, and America. It is an annual, and fends forth an irregular, branched, ligneous, hollow ftalk, which rifes about two feet high, and is furnifhed with oblong-oval, woolly leaves, on long downy footftalks. The flowers come out fingly from the fides of the branches, on long peduncles; thefe are fhaped like thofe of the common Potatoe, but their calyces are fet with fpines. They are fucceeded by large egg-fhaped berries, which are moftly of a purple colour on one fide, and white on the other. This plant varies much in the form and colour of its fruit, they being conical or egg-fhaped in fome, and in refpect to colour, are fometimes purple, pale red, yellow, or white. The plant is now frequently raifed in our

P 3   gardens,

gardens, where the fruit for the moſt part come white, and reſemble eggs, which has obtained it the name of *Egg Plant*. In the Weſt-Indies they call it *Brown John*, or *Brown Jolly*.

Theſe berries are boiled in ſoups and ſauces, the ſame as the Love Apple, are accounted very nutritive, and are much ſought after by the votaries of Venus.

38 SOLANUM ſanctum. *Paleſtine Night-ſhade. Lin. Sp. pl.* 269.

Solanum ſpinoſum, fructu rotundo. *Bauh. Pin.* 167.

This is a ſhrubby plant, and grows naturally in Egypt and Paleſtine. It hath a woolly, aſh-coloured ſtalk; ſet with ſhort, erect, thick, yellowiſh ſpines. The leaves are egg-ſhaped, and have ſerpentine edges; theſe are ſpiny and woolly. The flowers come out at the ſide of the ſtalks, on prickly peduncles; they are of a deep blue, with briſtly calyces, and they have a great reſemblance to the flowers of the Borrage.

Haſſelquiſt ſays this plant is known in Egypt, by the name *Meringam*, and that the fruit, which are globular, are much eaten by the inhabitants.

39 SORBUS domeſtica. *Lin. Sp. pl.* 684.

Sorbus ſativa. *Bauh. Pin.* 415.

The *True Service-tree* grows wild in the
warmer

warmer parts of Europe, and it is also found
in Cornwall, but many doubt its being a
native of England. It becomes a large tree,
sending out many branches, covered with a
rough grayish bark, and furnished with
winged leaves, resembling those of the common Ash, but they are hoary underneath,
(in the young trees), and serrated on their
edges. The flowers are produced in large,
round bunches at the ends of the branches;
they are small and white, consist of five petals each, surrounding many stamina, and
three styles. The germen is seated under
the flower, and becomes a soft, umbilicated
berry, inclosing three or four oblong, carti-
laginous seeds.

The natural size of these berries is that
of a small Medlar, but cultivation has alter-
ed both size and form; some being nearly
round, and as big as a Pippin, and others
Pear-shaped. They have a rough, astringent
taste when fresh gathered, and therefore
must be kept some time to mellow, and
then they become pleasant.

40 TROPHIS americana. *Red-fruited
Bucephalon. Lin. Sp. pl.* 1451.
Trophis foliis oblongo-ovatis glabris al-
ternis, floribus masculis spicatis ad alas.
Browne's Jam. 357.
This is a shrubby plant, and, as its tri-
vial name expresses, a native of America,

P 4   and

and particularly of the Ifland of Jamaica. It is male and female in diftinct plants. The leaves come out in an alternate order, on very fhort footftalks ; they are fmooth, of an oblong-oval form, fharp-pointed, and entire. The flowers are produced in long bunches, from the fides of the branches; thofe of the male have no calyx, but confift of four obtufe, fpreading petals, furround-ing four flender ftamina, longer than the petals. The female flowers are compofed of a fmall monophyllous calyx, and an oval germen, fupporting a bipartite ftyle; and are fucceeded by globular, rough berries, each having one cell, containing a roundifh feed.

Thefe fruits have not a very recommend-able flavour, yet are frequently eaten by the inhabitants of Jamaica.

41 VITIS vinifera. *Lin. Sp. pl.* 293.

The *Vine* is now multiplied into fo many varieties, that to fet them all down would be ufelefs, efpecially as feveral lifts of them have been already publifhed; but it will not be amifs, perhaps, to give fhort de-fcriptions of the few following, as they are in general efteem for their fuperior quali-ties, or are frequently cultivated in Eng-land. Thefe are :

1 The

1 The *Black Sweet Water*. 8 The *Black Muscat*.
2 The *White Sweet Water*. 9 The *Violet Muscat*.
3 The *Golden Chaffelas*. 10 *Alexandrian Muscat*.
4 The *Musky Chaffelas*. 11 The *Red and Black*
5 The *Black Cluster*.   *Hamburgh*.
6 The *White Muscat*. 12 *St. Peter's Grape*.
7 The *Red Muscat*.

The *Black Sweet Water* hath ſhort bunches, and ſmall roundiſh berries, growing cloſe together. Their ſkin is thin, and their juice very ſweet, which much tempt the birds and flies to deſtroy them. Ripens early in Auguſt.

The *White Sweet Water* hath very irregular ſized berries on the ſame bunch, ſome being of a good ſize, others extremely ſmall. The juice has a pleaſant ſugary flavour. It ripens with the former.

The *Golden Chaffelas* hath large bunches, and round, different ſized berries. Theſe arc of a bright green at firſt, and when ripe of an amber colour. The juice is ſweet and ſugary. The Red Chaffelas is a variety of this.

The *Musky Chaffelas* hath round berries, nearly of the ſize of the former. The berries are of a greeniſh-white, and plentifully ſtored with a ſugary, muſky, juice. It ripens in September.

The *Black Cluster* hath downy leaves, and ſhort bunches, cloſely ſet with oval berries, many of which cannot ripen, they being ſo covered with the reſt. This is by many called the *Burgundy*. Ripe about the time of the former.

The *White Muscat*, or *White Frontinac*, hath large, even, conical bunches, ending in a point. The berries are cloſely ſtudded together, and are of a bright green on the ſhady-ſide, inclining to

an

an amber colour on the other, and are thinly covered with a bloom. The juice has a moſt excellent flavour, when the berries are perfectly ripe,
which ſeldom happens here.

The *Red Muſcat*, or *Red Frontinac* hath long
bunches, more thinly ſet with berries than the
White. They are large and round; before ripe,
gray with dark ſtripes, but when fully ripe, almoſt of a brick red. The juice has a high, vinous flavour. Ripe the beginning of October.

The *Black Muſcat*, or *Black Frontinac*, hath good
ſized round berries, which are more diſtant on the
bunches than the Red. The bunches are ſhort,
the berries very black, and covered with a deep
violet bloom. The juice is very rich and vinous.
Ripe about the time of the former.

The *Violet Muſcat* hath leaves reſembling the
White Muſcat. The berries are large, rather
long, and are covered with a deep violet bloom.
The juice is not excellent, but muſky and agreeable.

The *Alexandrian Muſket*, or *Jeruſalem Muſcat*,
hath long, regular bunches, with the berries hanging looſe upon them. There are two ſorts of
this; one with whitiſh, and the other with red
berries, both of a rich, vinous flavour, but ſeldom ripen here.

The *Red and Black Hamburgh*, or *War-
ner's Grape*, has middle-ſized berries, and large
bunches. The former are rather of an oval ſhape,
and contain a ſugary, vinous juice. They ripen
in October.

The *St. Peter's Grape* hath very deep-divided
leaves, ſomewhat reſembling thoſe of the Parſley-
leaved Grape. The bunches are very large, the
berries

berries of a deep black, of an oval form, large, and make a fine appearance, but their juice is not rich.

The *Vine* is a native of France, Spain, Portugal, and many other places under the same parallels of latitude.

C H A P.

C H A P. VI.

ESCULENT STONE FRUIT*.

S E C T. I.

Stone Fruit of Europe.

1 AMYGDALUS perſica. *The Peach.*
        Nuciperſica. The Nectarine.
2 Cornus maſcula. *Male Cornel, or Corne-*
       lian Cherry.
3 Olea Europea. *Manured Olive.*
      —*ſylveſtris.* Wild Olive.
4 Prunus armeniaca. *The Apricot.*
5 Prunus avium. *Wild Black Cherry.*
6 Prunus ceraſus. *Wild Red Cherry.*
7 Prunus domeſtica. *The Plum-tree.*
8 Prunus inſititia. *The Bullace-tree.*
9 Rhamnus zizyphus. *Common Jujube.*

1 AMYGDALUS perſica. *The Peach.*
Lin. Sp. pl. 676.
  Perſica molli carne et vulgaris. *Bauh.*
Pin. 440.

---

* Linnæus defines *drupa* to be a pulpy pericarpium, or
feed-veſſel, without an opening, and includes a ſtone or
nut.

The

This is faid to be a native of Europe, but of what part is not known. The flower is compofed of five obtufe petals, inferted into a tubular calyx, cut into five obtufe fegments, together with above twenty flender ftamina, inferted alfo into the calyx, furrounding a roundifh germen, which turns to a roundifh, flefhy, furrowed fruit, inclofing a hard ftone. Cultivation has produced many varieties of this fruit, and the following are the moft efteemed forts.

1	The *White Nutmeg.*	15	The *Bellegarde.*
2	The *Red Nutmeg.*	16	The *Bourdine.*
3	The *Early Purple.*	17	The *Roffanna.*
4	The *Small Mignon.*	18	The *Admirable.*
5	The *White Magdalen.*	19	The *Old Newington.*
6	The *Yellow Alberge.*	20	The *Royal.*
7	The *Large French Mignon.*	21	The *Rambouillet.*
8	The *Beautiful Chevreufe.*	22	The *Portugal.*
		23	The *Late Admirable.*
9	The *Red Magdalen.*	24	The *Nivette.*
10	The *Chancellor.*	25	*Venus's Nipple.*
11	*Smith's Newington.*	26	The *Late Purple.*
12	The *Montauban.*	27	The *Perfique.*
13	The *Malta.*	28	The *Catharine.*
14	The *Vineufe.*	29	The *Monftrous Pavy.*
		30	The *Bloody Peach.*

The *White Nutmeg* is the firft Peach in feafon, it being often in perfection by the end of July. The leaves are doubly ferrated, the flower large, and of a pale colour; the fruit is white, fmall, and round; the flefh too is white, parts from the ftone, and has a fugary, mufky flavour.

The

The *Red Nutmeg* hath yellowifh green leaves, with ferpentine edges, which are flightly ferrated. The flowers are large, open, and of a deep blufh-colour. The fruit is larger, and rounder than the former, and is of a bright vermilion next the fun, but more yellow on the other fide. The flefh is white, except next the ftone, from which it feparates, and has a rich, mufky flavour. It ripens juft after the White Nutmeg.

The *Early Purple* hath fmooth leaves, terminated in a fharp point. The flowers are large, open, and of a lively red. The fruit is large, round, and covered with a fine deep red coloured down. The flefh is white, red next the ftone, and full of a rich, vinous juice. Ripe about the middle of Auguft.

The *Small Mignon* hath leaves flightly ferrated, and the flowers fmall and contracted. The Peach is round, of a middling fize, tinged with darkifh red on the fun-fide, and is of a pale yellowifh colour on the other. The flefh is white, parts from the ftone, where it is red, and contains plenty of a vinous, fugary juice. Ripens rather before the former.

The *White Magdalen* hath long, fhining, pale-green leaves, deeply ferrated on the edges, and the wood is moftly black at the pith. The flowers are large and open, appear early, and are of a pale red. The fruit is round, rather large, of a yellowifh-white colour, except on the funny fide, where it is flightly ftreaked with red. The flefh is white to the ftone, from which it feparates, and the juice is pretty well flavoured. Ripe at the end of Auguft.

The *Yellow Alberge* hath deep red, middle-fized flowers; the Peach is fmaller than the former, of

a yellow

a yellow colour on the shady side, and of a deep red on the other. The flesh is yellow, red at the stone, and the juice is sugary and vinous.

The *Great French Mignon* hath large, finely serrated leaves, and beautiful red flowers. The fruit is large, quite round, covered with a fine sattiny down, of a brownish red colour on the sunny side, and of a greenish yellow on the other. The flesh is white, easily parts from the skin, and is copiously stored with a sugary, high flavoured juice. Ripe near the middle of August.

The *Beautiful Chevreuse* hath plain leaves, and small contracted flowers. The fruit is rather oblong, of a middling size, of a fine red colour next the sun, but yellow on the other side. The flesh is yellowish, parts from the stone, and is full of a rich sugary juice. It ripens a little after the former.

The *Red Magdalen* hath deeply serrated leaves, and large open flowers. The fruit is large, round, and of a fine red next the sun. The flesh is firm, white, separates from the stone, where it is very red; the juice is sugary, and of an exquisite rich flavour. Ripe at the end of August.

The *Chancellor* hath large, slightly serrated leaves. The Peach is about the size of the Beautiful Chevreuse, but rather rounder. The skin is very thin, of a fine red on the sunny-side; the flesh is white and melting, parts from the stone, and the juice is very rich and sugary. It ripens with the former.

The leaves of *Smith's Newington* are serrated, and the flowers are large and open. The fruit is of a middle size, of a fine red on the sunny side; the flesh white and firm, but very red at the stone,

to

to which it sticks closely, and the juice has a pretty good flavour. Ripens with the former.

The *Montauban* hath serrated leaves, and large open flowers. The fruit is about the size of the former, of a purplish red next the sun, but of a pale one on the shady side. The flesh is melting, and white even to the stone, from which it separates. The juice is rich, and well flavoured. It ripens a little before the former.

The *Malta* hath deeply serrated leaves, and the flowers are large and open. The fruit is almost round, of a fine red next the sun, marbled with a deeper red, but the shady-side is of a deep green. The flesh is fine, white, except at the stone, from which it parts, where it is of a deep red; the juice is a little musky, and agreeable. It ripens at the end of August, or beginning of September.

The *Vineuse* hath large, deep green leaves, and full bright red flowers. The fruit is round, of a middle size; the skin is thin, all over red; the flesh fine and white, except at the stone, where it is very red, and the juice is copious and vinous. Ripe in the middle of September.

The *Bellegarde* hath smooth leaves, and small contracted flowers. The fruit is very large, round, and of a deep purple colour next the sun. The flesh is white, parts from the stone, where it is of a deep red, and the juice is rich and excellent. It ripens early in September.

The *Bourdine* hath large, fine green, plain leaves, and small flesh-coloured contracted flowers. The fruit is round, of a dark red next the sun, the flesh white, except at the stone, where it is of a deep red, and the juice is rich and vinous. Ripens with the former.

The

The *Roffanna* hath plain leaves, and fmall contracted flowers. The fruit is rather longer than the Alberge, and fome count it only a variety of the latter. The flefh is yellow, and parts from the ftone, where it is red; the juice is rich and vinous. Ripe early in September.

The *Admirable* hath plain leaves, and fmall contracted flowers, which are of a pale red. The fruit is very large and round; the flefh is firm, melting, and white, parts from the ftone, and is there red; and the juice has a fweet, fugary, high vinous flavour. Ripe early in September.

The *Old Newington* hath ferrated leaves, and large open flowers. The fruit is large, of a fine red next the fun; the flefh is white, fticks clofe to the ftone, where it is of a deep red, and the juice has an excellent flavour. It ripens juft after the former.

The *Royal* hath plain leaves, and fmall contracted flowers. The fruit is about the fize of the *Admirable*, and refembles it, except that it has fometimes a few knobs or warts. The flefh is white, melting, and full of a rich juice; it parts from the ftone, and is there of a deep red. Ripe about the middle of September.

The *Rambouillet* hath leaves and flowers like the Royal. The fruit is rather round than long, of a middling fize, and deeply divided by a furrow. It is of a bright yellow on the fhady-fide, but of a fine red on the other. The flefh is melting, yellow, parts from the ftone, where it is of a deep red, and the juice is rich and vinous. Ripe with the former.

The *Portugal* hath plain leaves, and large open flowers. The fruit is large, fpotted, and of a

Q   beautiful

beautiful red on the funny fide. The flefh is firm, white, fticks to the ftone, and is there red. The ftone is fmall, deeply furrowed, and the juice is rich and fugary. Ripe towards the end of September.

The *Late Admirable* hath ferrated leaves, and brownifh red fmall contracted flowers. The fruit is rather large and round, of a bright red next the fun, marbled with a deeper. The flefh is of a greenifh-white, and fticks to the ftone, where it hath feveral red veins; the juice is rich and vinous. Ripe about the middle of September.

The *Nivette* hath ferrated leaves, and fmall contracted flowers. The fruit is large and roundifh, of a bright red colour next the fun, but of a pale yellow on the fhady-fide. The flefh is of a greenifh-yellow, parts from the ftone, where it is very red, and is copioufly ftored with a rich juice. It ripens about the middle of September.

Venus's Nipple hath finely ferrated leaves, and rofe-coloured, fmall contracted flowers, edged with carmine. The fruit is of a middling fize, and has a rifing like a breaft. It is of a faint red on the fun-fide, and on the fhady one of a ftraw-colour. The flefh is melting, white, feparates from the ftone, where it is red, and the juice is rich and fugary. Ripens late in September.

The *Late Purple* hath large, ferrated leaves, which are varioufly contorted, and the flowers are fmall and contracted. The fruit is round, large, of a dark red on the funny-fide, and yellowifh on the other. The flefh is melting, white, parts from the ftone, where it is red, and the juice is fweet and high flavoured. Ripens with the former.

The *Perfique* hath large, very long indented leaves, and fmall contracted flowers. The fruit is large, oblong, of a fine red next the fun; the flefh firm, white, but red at the ftone, juicy, and of a high pleafant flavour. The ftalk has frequently a fmall knot upon it. Ripe late in September.

The *Catharine* hath plain leaves, and fmall flowers. The fruit is large, round, of a very dark red next the fun. The flefh white, firm, fticks clofe to the ftone, and is there of a deep red. The juice is rich and pleafant. It ripens early in October.

The *Monftrous Pavy* hath large, very flightly ferrated leaves, and large, but rather contracted flowers. The fruit is round, and very large, whence its name. It is of a fine red on the funny fide, and of a greenifh-white on the other. The flefh is white, melting, fticks clofe to the ftone, and is there of a deep red. It is pretty full of juice, which in dry feafons, is fugary, vinous and agreeable. Ripe towards the end of October.

The *Bloody Peach* hath rather large, ferrated leaves, which turn red in Autumn. The fruit is of a middling fize, the fkin all over of a dull red, and the flefh is red down to the ftone. The fruit is but dry, and the juice rather fharp and bitterifh. It feldom ripens well in England, but is well worth cultivating notwithftanding, for the fruit bake and preferve excellent well.

NECTARINES.

Linnæus makes the *Nectarine* only a variety of the *Peach*, for its having a fmooth coat was only an accident originally There are many varieties of it now cultivated; and

Q 2   the

the following are fome of the moft efteemed forts commonly planted in England.

1 The *Elruge.*   5 The *Murrey.*
2 The *Newington.*   6 The *Italian.*
3 The *Scarlet.*   7 The *Golden.*
4 The *Roman.*   8 The *Temple's.*

The *Elruge* hath large ferrated leaves, and fmall flowers. The fruit is of a middling fize, of a dark purple colour next the fun, and of a greenifh yellow on the fhady fide. The flefh parts from the ftone, and has a foft, melting, good flavoured juice. Ripe early in Auguft.

The *Newington* hath ferrated leaves, and large open flowers. The fruit is pretty large, of a beautiful red on the funny fide, but of a bright yellow on the other. The flefh fticks to the ftone, is there of a deep red colour, and the juice has an excellent rich flavour. Ripe towards the end of Auguft.

The *Scarlet* is rather lefs than the former, of a fine fcarlet colour next the fun, but fades to a pale red on the fhady fide. It ripens near the time of the former.

The *Roman,* or *Clufter Red Nectarine*, hath plain leaves, and large flowers. The fruit is large, of a deep red towards the fun, but yellowifh on the fhady fide. The flefh is firm, fticks to the ftone, and is there red; the juice is rich, and has an excellent flavour. Ripe about the end of Auguft.

The *Murrey* is a middling-fized fruit, of a dirty red colour on the funny-fide, and yellowifh on the fhady one. The flefh is firm, and tolerably well flavoured. It ripens early in September.

The *Italian Nectarine* hath fmooth leaves and
5   fmall

finall flowers; the fruit is red next the fun, but yellowifh on the other fide; flefh firm, adheres to the ftone, where it is red, and when ripe, which is early in September, has an excellent flavour.

The *Golden Nectarine* has an agreeable red colour next the fun, bright yellow on the oppofite fide; flefh very yellow, fticks to the ftone, where it is of a pale red, has a rich flavour, and ripens in September.

Temple's Nectarine is of a middling fize, of a fair red next the fun, of a yellowifh green on the other fide; flefh white near the ftone, from which it feparates; ripens in September, and has a high poignant flavour.

Peaches and *Nectarines* are wholefome fruits, and gently conftringe the ftomach, if eaten when not too mellow. The flowers of the former furnifh the fhops with an excellent fyrup for children, to whom it proves both gently emetic and cathartic.

2 CORNUS mafcula.   *Cornelian Cherry.* *Lin. Sp. pl.* 171.

Cornus fylveftris mas.   *Bauh. Pin.* 447.

This grows wild in the woods and hedges in Auftria. It is a fhrubby plant, dividing into many irregular branches, covered with a rough bark; thefe fpread wide, and are furnifhed with oval, veined leaves, not indented on their edges, and are fharp-pointed. The flowers come out in the fpring before the leaves, and at the ends of the branches, in diftinct umbels; they are fmall, yellowifh, compofed of four petals each, with four

Q 3   ftyle.

ftamina longer than the petals, and one ftyle. The germen is round, feated below the flower, and fwells to an umbilicated oval berry, containing a nut with two cells.

Thefe fruits are about the fize of Cherries, of a yellowifh red colour, and an auftere flavour, are therefore feldom eaten frefh off the bufhes, but are preferved to make tarts and other devices. There is a variety of this fhrub with white fruit.

3 Olea europea. *Manured Olive.* Lin. *Sp. pl.* 11.

Olea fativa. *Bauh. Pin.* 472.

This is an evergreen, and a native of Auftria, but is cultivated in France, Spain, Portugal, and Italy. It is rather of a fhrubby nature, frequently fending forth feveral ftems from the fame root, though fometimes there is only one. The branches are round-ifh, and furnifhed with fpear-fhaped leaves, of a bright green colour, and ftand oppofite. The flowers are produced in fmall bunches at the footftalks of the leaves; they are white, tubular, and cut into four fegments at the brims. Each flower contains two ftamina, which are much fhorter than the petal, and one flender ftyle, crowned with a fimple ftigma. The germen is roundifh, and turns to an oval plum, about the fize of a pigeon's egg, and when ripe of a greenifh black colour.

Thefe plums are pickled, and fent to different

ferent parts of Europe; but they are a very indifferent condiment, the oil with which they abound, being apt to pall and relax the ftomach. They vary very much in regard to their nature, fize, and colour, and this according to the foil and climate the trees are planted in. Thofe raifed in Italy are the fmalleft, have almoft an infipid tafte, and therefore are worth little. Thofe propagated in Spain are the largeft, but they have a rank, difagreeable fmell and flavour. The Provence Olives are of a fize between the two former, have a pleafant tafte, furnifh the moft efteemed oil, and are the moft valuable when pickled.

The greateft advantage arifing from the cultivated *Olive*, is the abundance of oil that is expreffed from the fruit; and this oil is of three forts. The pureft and moft valuable is that which runs upon a flight preffure; the next in goodnefs from the fame *Olives* more ftrongly preffed and flightly heated; and the laft and worft from the fame operation more forcibly repeated. The great utility of this oil is fufficiently known.

4 PRUNUS armeniaca. *The Apricot. Lin. Sp. pl.* 679.

Mala armeniaca majora. *Bauh. Pin.* 442.

In what particular part this grows naturally is not known. It rifes to a large tree,

Q 4   with

with wide extending branches, furnished
with nearly heart-shaped leaves. The flowers
have very short peduncles, and are composed
of five roundish petals, surrounding twenty
or more stamina, and one style. The varieties of this fruit most generally brought
to the table, are,

1 The *Masculine*.	5 The *Turkey*.
2 The *Orange*.	6 The *Alberge*.
3 The *Algier*.	7 The *Breda*.
4 The *Roman*.	8 The *Brussels*.

The *Masculine* is a small, roundish *Apricot*, red
on the sunny side, and of a greenish yellow on the
other. It puts forth a prodigious number of
flowers, and is the first ripe of any.

The *Orange* is a larger fruit than the former,
and when ripe of a deep yellow colour. The
flesh of this is not delicate, and therefore it is
more generally used for tarts.

The *Algier* is of an oval form, a little compressed on the sides, and of a pale yellow, or straw
colour when ripe. The flesh is dry, and but badly
flavoured.

The *Roman* is a larger fruit than the former,
and not quite so much flatted. The colour is
rather deeper, but the flesh is not so dry, and
better flavoured.

The *Turkey Apricot* is round, and larger than
either of the former. The flesh too is firmer, and
of a finer flavour.

The *Alberge* is a small, compressed fruit, of a
yellow colour on the sunny side, running into a
greenish yellow on the other.

The

The *Breda* is the beſt fruit of all the ſorts. It is large, roundiſh, of a deep yellow colour on the outſide, and of a gold colour within. The fleſh is ſoft, and full of a high flavoured juice. The ſtone is larger and rounder than in the others.

The *Bruſſels* is a middling ſized fruit, and ſomewhat of an oval form. The ſide next the ſun is red, with many dark ſpots ; but on the ſhady ſide it is of a greeniſh yellow. The fleſh is firm, and of a very good flavour. It is the lateſt ripe of all the Apricots.

5 Prunus avium. *Wild Black Cherry.*

Prunus umbellis ſeſſilibus, foliis ovato-lanceolatis conduplicatis ſubtus pubeſcentibus. *Lin. Sp. pl.* 680.

This grows wild in the woods of England, where it arrives to a very large tree, ſending out many ſpreading branches, the twigs of which are furniſhed with cluſters of oval, ſerrated leaves, ending in a plain, ſpear-ſhaped point, and ſupported by purpliſh footſtalks, having two linear, toothed ſtipulæ, or leaves at the baſe of each. The leaves are downy on the underſide, with many prominent ribs running almoſt to the margin. The flowers are produced in ſeſſile umbels, on long purpliſh peduncles, and for the moſt part come out by threes from the centre of ſeveral ſmall, ſcaly, oval *, concave leaves, having their upper ſurfaces

* Some of theſe are often cut into three lobes, both in this and the following ſpecies.

covered

covered with ſhort hairs, after the manner of the leaves of the *Sundew*. Theſe ſerve as an involucrum to the umbel. Each flower is compoſed of five white, oblong, ſnipped petals, inſerted into a ſmall ſmooth calyx, conſiſting of five acute ſegments, which turn back to the peduncle, and are of a bright purple colour at the inſertion of the petals. The fruits are ſmall, nearly egg-ſhaped, almoſt black when ripe, and contain a thick, ſweet juice, which greatly tempt the birds to deſtroy them. Theſe fruits are much uſed for making Cherry Brandy.

There is a ſort growing in ſome of the woods in Norfolk, which appears to be a variety of this; its leaves are ſmaller than the above, more finely ſerrated, are not quite ſo downy underneath, but the ſtipulæ and leaves of the involucrum are of the ſame form, and the inſides of the latter are equally hairy. The flowers are large, the fruit ſmall, red, egg-ſhaped, and bitteriſh.

The nurſerymen ſow the ſtones of the *avium* for raiſing ſtocks to graft or bud the other ſorts of Cherries upon; and the general opinion is, that the only garden-variety procured by ſowing the ſtones, is the *Black Corone*.

There is a water kept in the ſhops made from the fruit of the *Wild Black Cherry*, and has long been in much eſteem among nurſes as a remedy for convulſions in children, but

it

it is with good reafon now almoft laid afide;
for it has been proved, that the diftilled
water made from the ftones of thefe fruits
will poifon brutes very fuddenly, and as the
fhop water muft imbibe fome of the pernicious quality of the ftones, though probably in a fmall degree, yet the quantity
may be fufficient to hurt the tender nerves
of infants, and thereby increafe the diforder
it was intended to cure.

6 PRUNUS cerafus. *Wild Red Cherry.*
Prunus umbellis fubfeffilibus, foliis ovato-
lanceolatis conduplicatis glabris. *Lin. Sp.*
pl. 679.
This too grows in our woods and hedges,
is a much fmaller tree than the former, and
the bunches of flowers and leaves are fupported on fhort woody footftalks. The
leaves are but little better than half the fize
of thofe of the *avium*, more acute towards
the footftalk, and are fmooth and gloffy on
the underfide, the ribs are lefs prominent,
but they are ftudded with a few whitifh
hairs. The flowers are moftly produced
four or five together; their peduncles are
fmooth, fhort, and of a fhining green. The
fegments of the calyx are obtufe, the petals
roundifh, and very feldom fnipped. The
leaves of the involucrum are fhort, polifhed
on the outfide, and very flightly hairy on
the inner. The fruits are round, red, tolerably

lerably large, and of an acid flavour. Mr.
Hudſon now makes the *avium* only a variety
of this, but whoever will attend to the deſcriptions juſt given, will certainly conclude
he is wrong, and be fully convinced they
are diſtinct ſpecies.

Linnæus and other late writers on botany
have ſuppoſed the *ceraſus* to be the parent
of all the cultivated Cherries, except the
Black Corone; what induced them to con-
jecture this is difficult to gueſs, as ſeveral of
the garden ſorts retain more of the original
properties of the *avium*, than they do of the
ceraſus; and particularly the Bleeding Heart,
the White Heart, the Black Heart, and the
Ox Heart, the leaves, flowers, and involucra
of which differ but very little from thoſe of
the *avium* in its wild ſtate. Whether ſoil,
ſituation, or their being conſtantly budded
upon ſtocks raiſed from the ſtones of the
latter, have any ſhare in producing theſe
ſimilitudes, is uncertain, but if they be
diſtinct ſpecies, why ſhould not the one be
as liable to produce varieties as the other?
The following are the names of the ſorts
commonly cultivated in England.

1 The *Early May Cherry*.	7 The *White Heart*.
2 The *May Duke*.	8 The *Black Heart*.
3 The *Archduke*.	9 The *Red Heart*.
4 *Holman's Duke*.	10 The *Ox Heart*.
5 The *White Spaniſh*.	11 The *Bleeding Heart*.
6 The *Yellow Spaniſh*.	12 *Harriſon's Heart*.

13 *Tradeſcant's*

13 *Tradefcant's Cherry.* 17 The *Black Corone.*
14 The *Late Archduke.*   18 The *Large Mazard.*
15 The *Lukeward.*   19 The *Carnation.*
16 The *Red*, or *Kentifh.* 20 The *Morello.*

The fruit of moft of thefe varieties are well known, and therefore I fhall omit their particular defcriptions.

7 PRUNUS domeftica. *The Plum - tree.* *Lin. Sp. pl.* 680.

Prunus inermis, foliis lanceolato-ovatis. *Hort. cliff.* 186.

This grows wild in our woods and hedges. It is a fmaller tree than the former. The leaves are oval, and fpear-pointed. The flowers moftly ftand fingly, and the branches have no fpines. The cultivated varieties are many, and fome of them have a moft excellent flavour, but are deemed not very wholefome, and ought to be eaten fparingly. The following are fome of the moft efteemed forts; viz.

1 The *White Primordian.* 9 The *White Perdigron.*
2 The *Early Black Da-*   10 The *Bonum-magnum.*
   mafk.   11 The *White Mogul.*
3 The *Little Black Da-* 12 The *Chefton.*
   mafk.   13 The *Apricot Plum.*
4 The *Great Damafk Vi-* 14 The *Maître Claude.*
   olet.   15 The *Red Diaper.*
5 The *Fotheringham.*   16 The *Small Queen*
6 The *Orleans.*   *Claude.*
7 The *Black Perdigron.* 17 The *Large Queen*
8 The *Violet Perdigron.*   *Claude.*

18 The

18 The *Myrobalan.*
19 The *Date Plum.*
20 The *Cloth of Gold.*
21 The *St. Catharine.*
22 The *Royal Plum.*
23 The *Brignole.*
24 The *Emprefs.*

25 The *Late Red Da-*
   mafk.
26 The *Wentworth.*
27 The *Bricette.*
28 The *White Pear Plum.*
29 The *Mufcle Plum.*
30 The *St. Julian.*

The *White Primordian* is a yellow, fmall, longifh Plum, covered with a white flue. It is but an indifferent fruit, and has only its earlinefs to recommend it, being ripe by the middle of July.

The *Early Black Damafk* is a round, middling-fized Plum, divided with a furrow, and is of a dark black colour, covered with a violet flue. The flefh is yellow, of a good flavour, and parts from the ftone. Ripe the beginning of Auguft.

The *Little Black Damafk* ripens juft after the former. It is fmall, and covered with a light violet bloom ; the flefh parts from the ftone, and has a fweet, fugàry juice.

The *Great Damafk Violet* is inclining to an oval fhape. The fkin is of a dark blue, covered with a violet bloom. The flefh is yellow, parts from the ftone, and the juice is richly fugared. Ripe in the middle of Auguft.

The *Fotheringham* is of a blackifh red colour, is rather of an oblong form, and deeply furrowed in the middle. The flefh is firm, parts from the ftone, and the juice is very rich. Ripe with the former.

The *Orleans* is a round, middle-fized Plum, of a blackifh red colour on the outfide, and of a yellowifh green within. The flefh is firm, parts from the ftone, and has a tolerable good flavour. Ripe with the former.

The

The *Black Perdigron* is an oval, middle-fized Plum, of a very dark colour, covered with a violet bloom. The flesh is firm, and copioufly ftored with an excellent rich juice. Ripe at the end of Auguft.

The *Violet Perdigron* is a large, roundifh Plum, of a bluifh-red colour on the outfide. The flesh is yellowifh, fticks to the ftone, and the juice has a moft exquifite rich flavour. Ripe with the former.

The *White Perdigron* is an oval, middling-fized fruit, of a yellow colour, covered with a white bloom. The flesh is firm, and has an agreeable fweetnefs. Ripe the end of Auguft.

The *Red Bonum-magnum* is a large, deep-red, oval Plum, covered with a fine bloom. The flesh is firm, fticks to the ftone, and has an auftere, acid flavour, on which account it is moftly ufed for tarts. Ripe in September.

The *White Mogul* is alfo a large, oval fruit, of a yellowifh colour, covered with a white bloom. The flesh is acid, and unpleafant raw, but it bakes well. Ripe juft after the former.

The *Chefton* is an oval, middle-fized Plum, of a dark blue colour, powdered with a violet bloom. The juice is rich, and it is a great bearer. Ripe about the middle of September.

The *Apricot Plum* is large, round, and yellow, and is covered with a white bloom. The flesh is firm, parts from the ftone, and has a fweet flavour. Ripe foon after the former.

The *Maître Claude*, as it is called in England, is a middle-fized Plum, of a fine mixed colour, between red and yellow, and is of a roundifh figure. The flesh is firm, parts from the ftone, and has a good flavour. Ripe in September.

The

The *Red Diaper* is a large, round Plum, of a reddish colour, covered with a violet bloom. The flesh has a very high flavour, and sticks to the stone. Ripe about the middle of September.

The *Small Queen Claude* is a round, whitish-yellow Plum, covered with a pearl-coloured bloom. The flesh is thick, firm, parts from the stone, and the juice is richly sugared. Ripe with the former.

The *Large Queen Claude* is a middling-sized, round, yellowish green fruit. The flesh is firm, of a dark green colour, parts from the stone, and the juice has an exceeding rich flavour. This is often confounded with the Green Gage, but it is a better Plum. Ripe about the middle of September.

The *Myrobalan* is a round, middle-sized Plum, of a dark purple colour, powdered with a violet bloom. The juice is sweet, and it is ripe early in September.

The *Date Plum* too is of a middle-size, but rather inclining to oblong. The skin is of a fine yellow, and frequently marked with bright red spots. The shady side is green, with a white bloom. Ripe in September.

The *Cloth of Gold* is a rounder Plum than the former, and more streaked with red. The flesh is yellow, and full of an excellent rich juice. Ripe about the middle of September.

The *St. Catharine* is an oval fruit, a little flatted. The skin is of an amber colour, covered with a whitish bloom; but the flesh is of a bright yellow, firm, sticks to the stone, and has an agreeable, sweet flavour. Ripe just after the former.

The *Royal Plum* is a large, oval fruit, and pointed at the stalk. It is of a light red colour, covered with a whitish bloom. The flesh sticks

to

to the ftone, and has a pleafant, fugary juice. Ripe towards the end of September.

The *Brignole* is a large, oval Plum, of a yellowifh colour, mixed with red. The flefh is of a bright yellow, is dry, but of an excellent tafte. Ripe about the middle of September.

The *Emprefs* is rather a large, oval Plum, of a violet colour, and thickly covered with a whitifh bloom. The flefh is yellow, fticks to the ftone, and has a very agreeable flavour. Ripe at the end of September.

The *Late Red Damafk* is a middling-fized Plum, of an oval form. It is of a deep red on the funny-fide, and of a pale one on the other. The flefh is yellowifh, melting, and of a good flavour. Ripe late in September.

The *Wentworth* is a large, oval Plum, of a yellow colour, and much refembles the *Bonum-magnum*. The flefh is yellow, parts from the ftone, and has a fharp, acid tafte. It ripens at the end of September, and is principally ufed for tarts.

The *Bricette* is a fmall, yellowifh-green Plum, powdered with a white bloom. The flefh is yellow, fweet, but of a flattifh flavour. Ripe the beginning of October.

The *White Pear Plum* is a rather longifh, white fruit, of an unpleafant, acid flavour, and therefore not proper to eat raw, but is a good fruit for preferving. It comes fo late that it feldom ripens well.

The *Mufcle* is an oblong, pointed Plum, of a dark blue colour. The ftone is large, and the flefh thin. There are feveral forts of the Mufcle Plum, as the Black, the Red, and the White, but they have all but an indelicate flavour.

The *St. Julian* is a fmall, dark violet-coloured

R   Plum,

Plum, covered with a mealy bloom. The flesh
sticks to the stone, and in fine autumns the fruit
will dry upon the trees. These last three sorts
are raised more for stocks to bud upon, than for
their fruits.

8 PRUNUS infititia. *The Bullace - tree.*
Lin. Sp. pl. 680.

Pruna fylveftria præcocia. *Bauh. Pin.*
444.

This grows wild in our hedges. The
flowers are moftly produced two together.
The leaves are more oval than thofe of the
domeftica, are downy underneath, and the
edges are rolled inward. The branches are
a little fpiny. The *Black Bullace* is too
well known to require a defcription. There
are two varieties of it, the Red and the
White Bullace.

9 RHAMNUS zizyphus, *Common Ju-*
jube. Lin. Sp. pl. 282.

Jujuba fylveftris. *Bauh. Pin.* 446.

The *Common Jujube* is a native of the
warm parts of Europe. It hath a ftiff
woody ftem, which divides into many irregular branches, fet with erect fpines in
pairs. The leaves are of an oblong-oval
form, fmooth, and flightly ferrated on the
edges; they are about two inches long, and
ftand upon fhort footftalks. The flowers
are produced by two or three at a place;

5   are

are yellowifh, funnel-fhaped, have no calyx, and are cut into five fegments at their brims. Each includes five awl - fhaped ftamina, faftened to the bafe of the petal, and two flender ftyles, crowned with two obtufe ftigmata. The germen becomes an oval Plum, inclofing a ftone with two cells, each having an oblong feed.

The fruit are about the fize of Olives, of a yellowifh red colour, fweetifh, and a little clammy. In the winter feafon they are ferved up at table in Spain and Italy, as a dry fweetmeat. They were formerly kept in the fhops, by the name of Jujubes, and ftood recommended againft coughs, afthmas, pleurifies, and heat of urine; but are feldom to be met with at prefent.

SECT. II.

Stone Fruit Exotic.

1 CHryfobalanus icaco. *Cocoa Plum.*
2 Coccoloba uvifera. *Sea-fide Grape,* or *Sea-fide Mangrove.*
3 Cordia myxa. *Cluftered Sebeften,* or *Affyrian Plum.*
4 Cordia febeftena. *Rough-leaved Sebeften.*
5 Corypha umbraculifera. *Umbrella Palm.*
6 Elais guineenfis. *Oil Palm.*

R 2   7 Eugenia

7 Eugenia jambos. *Malabar Plum.*
8 Grias cauliflora. *Anchovy Pear.*
9 Laurus persea. *Avigato Pear.*
10 Mangifera indica. *Mango-tree.*
11 Phœnix dactylifera. *Common Date.*
12 Rhamnus jujuba. *Indian Jujube.*
13 Spondias lutea. *Yellow Jamaica Plum.*

1 CHRYSOBALANUS icaco. *Cocoa Plum.* *Lin. Sp. pl.* 681.

This tree is a native of South America, growing there in many parts near the sea. It is a shrubby plant, not rising more than eight or ten feet high, and sending out many side branches, covered with a dark brown bark, spotted with white; these are furnished with stiff, rough leaves, which are snipped at their ends into the form of an inverted heart, and stand in an alternate order on short footstalks. Both at the wings of the leaves, and divisions of the branches, the flowers are produced in loose panicles. They are small and white, consist of a bell-shaped calyx each, cut into five spreading parts at the brim, containing five oblong petals, inserted by their bases into the calyx. The stamina are ten, or more, tipped with yellow summits; these surround a long style, sitting upon an oval germen, and crowned with an obtuse stigma.

The fruit are about the size of small Olives, and of various colours, some being

3 whitish,

whitifh, fome brown, fome blue, and others
blackifh. The ftone is fhaped like a pear,
and has five longitudinal furrows. The
Plums have a fweet lufcious tafte, and are
brought to the tables of the inhabitants
where they grow, by whom they are much ·
efteemed.

2 Coccoloba uvifera. *Sea-fide Grape.*
Lin. Sp. pl. 523.

Populus americana rotundifolia. *Bauh.*
Pin. 430.

The *Sea-fide Grape* grows upon the fandy
fhores of moft of the Weft India iflands,
where it fends up many woody ftems, eight
or ten feet high, covered with a brown
fmooth bark, and furnifhed with thick,
veined, fhining orbicular leaves, five or fix
inches diameter, ftanding upon fhort footftalks. The flowers come out at the wings
of the ftalks, in racemi of five or fix inches
long; they are whitifh, have no petals, but
each is compofed of a monophyllous calyx,
cut at the brim into five dblong, obtufe fegments, which fpread open, continue, and
furround feven or eight awl-fhaped ftamina,
and three fhort ftyles, crowned with fimple
ftigmata. The germen is oval, and becomes
a flefhy fruit, wrapped round by the calyx,
and includes an oval nut, or ftone.

Thefe Plums are about the fize of Goofe-
berries, of a purple red colour, and a tole-

R 3   rable

rable good flavour. There are some other species of this genus whose fruits are eaten by the inhabitants where they grow, but they are smaller, and not so well tasted.

3 CORDIA myxa. *Affyrian Plum. Lin. Sp. pl.* 273..

Sebeftena fylveftris et domeftica. *Bauh. Pin.* 446.

The *Cultivated Sebeften* grows wild in Affyria and Egypt, and also on the coaft of Malabar. It rifes to the height of a middling Plum-tree, and its branches are furnished with oval, woolly leaves, ftanding without order. The flowers are produced in bunches, are white, and confift of one tubular petal, and a like calyx, nearly of an equal length, and both are cut into five parts at their brims. In their centre are five very fmall ftamina, and one flender ftyle, crowned with an obtufe ftigma. The germen is roundish, and fwells to a Plum of the fame form, and about the fize of a Damfon, of a dark brown colour, a fweet tafte, and very glutinous.

Thefe Plums were formerly kept in the fhops, and were accounted good for obtunding acrimony, and thereby ftopping defluxions of rheum upon the lungs; but at prefent they are little ufed for thefe purpofes.

In fome parts of Turky they cultivate
this

this tree in great abundance, not only for the fake of the fruit to eat, but to make bird-lime of, which is a vaft article of trade in a town called *Seid.*

4 CORDIA febeftena. *Rough-leaved Se-befien. Lin. Sp. pl.* 271.
Cordia foliis amplioribus hirtis, tubo floris fubæquali. *Browne's Jam.* 202.
This grows naturally in both the Indies, and fends forth feveral fhrubby ftalks eight or ten feet high. The young leaves are ferrated, but the full grown ones are not. They are of an oblong-oval form, rough, of a deep green' on the upper fide, and ftand alternately on fhort footftalks. The flowers terminate the branches in large clufters, are nearly of the fhape and colour of thofe of the Marvel of Peru, and make a moft beautiful appearance. Each has five ftamina, and one bifid ftyle. The Plums are much of the fhape of thofe of the *myxa,* and are eaten in the fame manner.
The fruit of this tree is lefs valuable than the wood, a fmall piece of which thrown upon a clear fire will perfume a room with a moft agreeable odour.

5 CORYPHA umbraculifera. *Umbrella Palm. Lin. Sp. pl.* 1657.
Palma montana, folio plicatili flabelli-
R 4   formi

formi maximo femel tantum frugifera.
Raii Hift. 1363.

This is a fpecies of *Palm*, and a native of
India, where it is called *Codda-pana*. It
rifes to a confiderable height, and produces
at the top many large palmated, plaited
leaves, the lobes of which are very long,
and are placed regularly round the end of a
long fpiny footftalk, in a manner reprefenting a large umbrella. The flowers are produced on a branched fpadix, from a compound fpatha or fheath; they are hermaphrodite, and each confifts of one petal, divided into three oval parts, and contains fix
awl-fhaped ftamina, furrounding a fhort
flender ftyle, crowned with a fimple ftigma. The germen is nearly round, and becomes a large globular fruit of one cell, including a large round ftone. Thefe Plums
having a pleafant flavour are held in efteem
by the Indians.

6 ELAIS guineenfis.   *Oil Palm.*   *Lin.*
Syft. Nat. 730.

Palma frondibus pinnatis ubique aculeatis
nigricantibus, fructu majore.   *Mill. Dict.*

This too is a fpecies of *Palm*, and grows
fpontaneoufly on the coaft of Guinea, but is
much cultivated in the Weft-Indies. It
rifes to forty or fifty feet high, bearing at
the top many winged leaves, the lobes of
which

which are long, narrow and flexible. The footstalks of the leaves clasp the stem with their broad bases, from which they regularly diminish upward, and are all the way set with strong, recurved, blackish spines. The flowers are male and female in separate bunches, and come out between the leaves; those of the male are monopetalous, cut at their brim into six segments, and each has a six-leaved calyx; in the centre are six slender stamina longer than the petal. The females have likewise a six-leaved calyx and six distinct petals, including three stigmata. The germen is oval, and swells to a fruit somewhat bigger than an Olive of a yellow colour, and contains a stone with three valves.

These fruits are copiously stored with a sweet luscious oil, which the Indians are very fond of, and their manner of extracting it, is to roast the fruit in the embers, and then suck the oil out of them. But for the purpose of keeping, they draw the oil in the same manner as the Europeans do that of Olives, and use it in diet as we do butter. It is of the consistence of an ointment, of an orange colour, a pleasant taste, of no disagreeable smell, and enters our *materia medica* as an emollient, and a strengthener of all kind of weakness of the limbs. It also stands recommended against bruises, strains, cramps, pains, swellings, &c.

The

The Indians anoint their bodies with this oil, not only to prevent a too plentiful per-fpiration, but to fupple their ftiffened fibres, and to render their fkins foft and fleek. The ftones of the fruit contain agreecable-flavoured kernels, which the Negroes fcoop out, and then ftring the fhells in the manner of beads, in order to wear about their necks. This is a valuable tree to the inhabitants, for befides the benefits already mentioned to accrue from it, they alfo extract a liquor from the body, which they ferment into an intoxicating drink, called *Palm-wine*.

7 Eugenia jambos.　*Malabar Plum.* Lin. Sp. pl. 672.

Perfici officulo fructus malaccenfis ex candido rubefcens. *Bauh. Pin.* 441.

This is a very tall tree, and a native of India. The body is covered with a greyifh bark, and it fends out many fpreading branches, in the manner of the Walnut. The leaves are oblong, entire, fharp-pointed, of a deep green on their upper fide, of a pale one underneath, and are five or fix inches long. The flowers come forth at the ends of the twigs, on branched peduncles. Each is compofed of a monophyllous calyx, cut into four obtufe fegments; and four oblong, obtufe petals, twice the length of the calyx, with many ftamina inferted into them. The germen is feated underneath;

underneath ; it is top-fhaped, fupports a
ftyle longer than the ftamina, and becomes
a fruit about the fize of a fmall Pear, having one cell, containing a roundifh ftone.

The fruit vary in their colour from a fiefh
to a dark red, and fmell like Rofes. On the
coaft of Malabar, where the trees grow plentifully, thefe plums are in great efteem.
They are not only eaten frefh off the trees,
but are preferved with fugar, in order to
have them at table at all times in the year.
Of the flowers they make a conferve, as we
do of Rofes, which is ufed medically for
the fame purpofes as the latter is.

8 Grias cauliflora. *Anchovy Pear. Lin.
Sp. pl.* 732.

Calophyllum foliis tripedalibus obovatis,
floribus per caulem et ramos fparfis. *Browne's
Jam.* 245.

The *Anchovy Pear* is a native of Jamaica. The leaves are nearly oval, and about
three feet long. It hath a ftraight ftem,
upon the upper part of which come forth
the flowers, each compofed of a monophyllous calyx, containing four roundifh, ftiff,
concave petals, and many briftly ftamina,
inferted into the calyx. The germen is depreffed, funk in the calyx, has no ftyle,
but fupports a crofs-fhaped ftigma. The
fruit is large, and contains a ftone with eight
furrows.

Thefe

Thefe fruits are eaten by the inhabitants, but their flavour or quality I know nothing of.

9 LAURUS perfea. *Avigato Pear. Lin. Sp. pl.* 529.

Pyro fimilis fructus in Nova Hifpania, nucleo magno. *Bauh. Pin.* 439.

The *Avigato Pear* is a native of the Weft-India Iflands, and is a large tree, growing thirty or forty feet high. The trunk is covered with a fmooth afh-coloured bark, and the branches are furnifhed with large leaves like thofe of Laurel, but of a tougher texture; thefe are of a deep green colour, and continue the year through. The flowers are moftly produced near the extremities of the branches; they are of a dirty yellow colour, and agreeable fmell, have no calyx, but each is compofed of fix oval, fharp-pointed, fpreading petals, furrounding nine ftamina, (three of which are often imperfect) about half the length of the petals, and one fhort ftyle. The germen is Pear-fhaped, and fwells to a large flefhy fruit of the fame form, covered with a ftrong, tough fkin or fhell, which is fmooth, of a beautiful green at firft, but when ripe of a yellow colour, and contains a pale green pulp, that melts in the mouth like marrow, which it greatly refembles in flavour, and is very nourifhing. Dr. Bancroft fays it is the moft
nutritious

nutritious of all the tropical fruits. Within is a large, roundifh, ruffet-coloured
wrinkled nut, without any kernel.

Though this tree is faid to be a native
of the Weft-Indies, yet it is probable it
was originally brought thither from New
Spain, where it grows in great abundance,
and is of great ufe to the inhabitants. The
unripe fruit have but little tafte, neverthelefs, they being very falubrious, and of a
refrefhing comfortable nature, are frequently
brought to table, and eaten with falt and
pepper. The failors, when they arrive at
the Havanna and thofe parts, purchafe
plenty of thefe fruits, and chopping them
into fmall pieces with green Capficums and
a little falt, regale themfelves moft heartily
with them.

As the pulp is very foft and delicious in
the ripe fruit, the inhabitants often break
the fhells and fcoop out the marrow with a
tea-fpoon ; but the moft common method
is to ferve it up to table on a plate, mixed
with fugar, rofe-water, and the juice of
Limes, which render it quite delicate, and
in this form it warms and fortifies the ftomach, and is counted good againft dyfenteries.

Of the buds of this tree a ptifan is made;
which is deemed excellent againft the vene-
real difeafe ; and an infufion of them, drank
in a morning fafting, is ftrongly recommended

mended for diſlodging coagulated blood in
the ſtomach, produced there by means of a
ſtroke or fall. The wild hogs greedily devour the fruit of this tree, and thoſe of the
Mammea, which give their fleſh a moſt
agreeable and luſcious flavour.

10 MANGIFERA indica. *Mango-tree.*
Lin. Sp. pl. 290.
Perſicæ ſimilis putamine villoſo. *Bauh.*
Pin. 440.
The *Mango-tree* grows naturally on the
coaſt of Malabar, but is cultivated almoſt
all over Aſia. It is a large ſpreading tree,
having the branches thickly ſet with long,
narrow leaves, ſomewhat reſembling thoſe
of the Peach, but larger. The flowers come
out in compound racemi, are compoſed of
five white, ſpear-ſhaped petals each, ſurrounding five awl-ſhaped ſtamina, longer
than the petals, and tipped with heart-
ſhaped ſummits. The germen is roundiſh,
ſupports one ſlender ſtyle, crowned with a
ſimple ſtigma, and ſwells to a kind of kidney-ſhaped fruit, about the ſize of a Peach,
and covered with a ſoft downy ſkin of like
nature.
Theſe fruits when ripe are juicy, of a good
flavour, and are ſo fragrant, as to perfume
the air to a conſiderable diſtance. They
are eaten either raw, or preſerved with ſugar. Their taſte is ſo luſcious that they
soon

foon pall the appetite. The unripe fruits
are pickled in the milk of the Cocoa Nut
that has ftood till four with falt, Capficum,
and garlick, and thus managed they are eat-
en in the manner of Mango, and are faid to
have a pleafant flavour.

11 PHOENIX dactylifera. *Common Date.*
Lin. Sp. pl. 1658.

Palma dactylifera major vulgaris. *Sloan.*
Jam. 174.

The *Date-tree* is a fpecies of *Palm*, and
grows plentifully in Africa and moft parts of
India. It hath a fort of pithy trunk, which
in fome places rifes to near an hundred feet.
This is round, ftraight, and ftudded with
protuberances, which are the veftiges of de-
cayed leaves; for as the tree advances in
height, the old leaves fall off. When the
tree is arrived to a bearing ftate, the leaves
at the top are fix or eight feet long, extending all round like an umbrella, and regularly bending towards the earth. They
are pinnated, with lobes near a yard long,
about an inch broad, fharpifh pointed, and
of a bright green colour. The trees are
male and female in diftinct plants. The
flowers of both come out between the
leaves; thofe of the male are produced on a
long branched fpadix, iffuing from a large
fpatha, and are compofed of a fmall tripar-
tite

tite * calyx, containing three oval, white petals, and three very short stamina, tipped with long, four-square summits. The female flowers come out in the same manner as the former, and much resemble them, but have a roundish germen, supporting a short style, crowned with an acute stigma. When these fall they are succeeded by fruit about the size of Olives, but of different casts and colours on the outside, and contain a yellowish, agreeable-flavoured pulp, in the midst of which is a round, hard stone, of an ash-colour, and marked with deep furrows.

Unripe *Dates* are rather rough and astringent, but when they are perfectly matured, they are much of the nature of the Fig. The Senegal *Dates* are deemed the best, they having a more sugary agreeable flavour than those produced at Egypt, and other places. This tree is of inestimable value to the inhabitants where it grows, almost every part serving some œconomical purpose. Dr. Hasselquist's relation of it is as follows:

" In Upper Egypt many families subsist almost entirely upon *Dates*; in Lower Egypt they do not eat so many; rather choosing to sell them. The Egyptians make a conserve with fresh *Dates*, mixing them

* Cut into three parts.

with

with fugar; this has an agreeable tafte.
The kernels of the *Dates* are as hard as
horn, and no one would imagine that any
animal would eat them. But the Egyptians break them, and grind them in their
mills, and, for want of better food, give
them to their Camels, who eat them. In
Barbary, they turn beads for pater-nofters,
of thefe ftones. Of the leaves they make
bafkets, or fhort bags, which are ufed in
Turkey, on journies, or in their houfes.
In Egypt they make fly-flaps of them, convenient enough to drive away thefe numerous infects, which much incommode a man
in this country. I have likewife feen brufhes
made of them, with which they clean their
foffas and cloaths. The hard boughs they
ufe for fences round their gardens, and
cages to keep their fowls in, with which
they carry on a great traffick. They alfo
ufe the boughs for other things in hufband-
ry, inftead of wood, which they are deftitute of. The trunk or ftem is fplit, and
ufed for the fame purpofes as the branches,;
they even ufe it for beams to build houfes,
as they are ftrong enough for fmall buildings.
It is likewife ufed for firing, where there is
want of better. The integument, which
covers the tree between the boughs, entirely
refembles a web, and has threads, which
run perpendicularly and horizontally over
one another; this is of confiderable ufe in

S　　　　　Egypt,

Egypt, for of it they make all the rope they use to their cisterns, &c. They have also rigging of the same kind for their smaller vessels; it is pretty strong and lasting. They reckon in Egypt, that *Date - trees* afford to their owners a Sequin * annually of profit for each tree. It is common to see two, three, or four hundred fruit-bearing trees all belonging to one family, and one may sometimes see three or four thousand in the possession of one man, which, at the above rate, bring in a considerable revenue to their owner, for the little spot of ground they occupy. A full grown *Date-tree* does not, at most, take up above four feet in diameter, so that they may be planted within eight feet of each other."

The *Date-tree*, as has been shewn in the description, is male and female in distinct plants, and the husbandry practised by the cultivators of these trees, in order to be sure of a crop, is one of the main pillars that support the sexual system; for, unless the flowers of the female be impregnated by those of the male, the crop will be very scanty, and the quality of the fruit inferior, nor will the stones of such *Dates* vegetate when sown. It greatly behoves

---

* A Sequin in Egypt is worth about nine shillings sterling, and allowing nine feet for every tree (which is one foot more than Hasselquist mentions) an acre of land would contain 1613 trees, and produce to the owner 725 pounds annually.

female

the hufbandman, therefore, to fee that his
female trees are plentifully fupplied with
the farina of the male, and as the manner
of performing this is curious, and may be
new to many readers, Dr. Haffelquift's relation of it may not prove unacceptable. In
a letter to Dr. Linnæus, dated at Alexandria.
———" The firft thing I did, fays he, after
my arrival in Egypt, was to fee the *Date-
tree*; the ornament, and a great part of the
riches of this country. It had already blof-
fomed, but I had, neverthelefs, the pleafure
of feeing in what manner the Arabs affift
its fecundation, which is as follows: when
the fpadix, or receptacle of the *Date*, bears
female flowers, they fearch on a male *Date-
tree* for a fpadix, which has not yet burft,
or been protruded from its fheath; this they
open, take out its fpadix, and cut it lengthways in feveral pieces, taking care not to
hurt the flowers; a piece of this fpadix
with male flowers, is put lengthways between the fmall branches of the fpadix with
female flowers, over which is laid a Date-
leaf. In this fituation I yet faw the greateft
part of the fpadices, or heads of flowers,
which bore their young fruit; but the
male flowers, which were intermingled
with the female, were withered. The Arab,
who informed me of thefe particulars, gave
me likewife the following anecdotes. Firft,
unlefs they wed, and fecundate the *Date-*

tree

tree in this manner, it bears no fruit *.
Secondly, they always take the precaution to
preferve fome unopened fpathæ with male
flowers, from one year to another, to be
applied for this purpofe, in cafe the male
flowers fhould mifcarry, or fuffer damage.
Thirdly, if they permit the fpadix of the
male flowers to burft, or come out, it becomes ufelefs for fecundation : it muft have
the] maidenhead, fay the Arabs, which is
loft in the fame moment the bloffoms burft
out of their cafe. The perfon, therefore,
who cultivates *Date-trees*, muft be careful
to hit the proper time of affifting their
fecundation, which is almoft the only article in their cultivation."

12 RHAMNUS jujuba. *Indian Jujube.*
Lin. Sp. pl. 282.

The *Indian Jujube* is a fmaller tree than
the *Zizyphus*, defcribed in the laft Sect.
The branches of this are covered with a
yellowifh bark, and the fpines are bent, and
ftand fingly, whereas thofe of the *Zizyphus*
are ftraight, and placed two together. The
leaves are almoft round, woolly underneath,

* This muft be underftood, that it bears no fruit of a
good quality, and fuch as the feeds will not vegetate when
fown, by reafon they want the *punctum vitæ*, the fame as
eggs laid without the affiftance of a cock; which, though
they may appear perfect in every refpect, yet wanting the
fpeck of life, can never be brought one jot the forwarder by
the incubation of the hen.

and notched at the footftalks. The flowers
come out in clufters, fome having two
ftyles, others only one. The fruit are almoft globular, and have been by many fuppofed to be the true Sebeften of the fhops,
but Linnæus and his difciples have amply
proved the contrary, and fhewn that the
fhop Sebeften is the fruit of the *Zizyphus*.

13 SPONDIAS lutea.   *Yellow Jamaica
Plum. Lin. Sp. pl.* 613.

Spondias foliis plurimis pinnatis ovatis,
racemis terminalibus, cortice interno rubente. *Browne's Jam.* 229.

This tree is a native of America, and it
is highly probable it grows alfo in the Eaft
Indies. It is of fmall ftature, feldom rifing
more than twelve or fourteen feet, breaking
into many branches, which are furnifhed
with pinnated leaves, compofed of a great
number of ferrated pinnæ, placed alternately
along the midrib, which is terminated by
an odd one. The flowers are produced at
the ends of the branches, in long racemi,
they are of a pale yellow colour, and each
confifts of a fort of bell-fhaped calyx, cut
into five fegments, together with five oblong,
plain, fpreading petals, furrounding ten
briftly ftamina, fhorter than the petals, and
five fhort, perpendicular ftyles, crowned
with obtufe ftigmata. The germen is oval,
and becomes an oblong fruit, of a pale

yellow

yellow colour, covered with a mealy farina, and contains a woody, fibrous ſtone, having five cells.

Theſe Plums have a ſweet luſcious taſte, but are thinly furniſhed with fleſh, other-wiſe they would be much more valued; they are, however, in general eſteem among the inhabitants of the Weſt India iſlands, and are of great uſe to the hogs, being their principal food all the time they are in ſeaſon.

It is probable theſe Plums were one of the ſorts of Myrobalans formerly kept in the ſhops, which conſiſted of five different ſpecies. There is another tree of this genus, natural to the Eaſt Indies, and differs little from this, but in the colour of the fruit, which is purple, and therefore it is not un-likely but this was another of the ſhop My-robalans, as one ſort of them was of this colour.

C H A P.

C H A P. VII.

ESCULENT APPLES*.

S E C T. I.

Apples of Herbaceous Plants.

1 CUCUMIS melo. *Muſk Melon.*
——— *melo albus.* Spaniſh White Melon.
——— *melo lævis.* Smooth, green-fleſhed Melon.
——— *melo flavus.* Yellow Winter Melon.
——— *melo parvus.* Small Portugal Muſk Melon.
——— *melo piloſus.* Hairy-ſkinned Melon.
——— *melo reticulatus.* Netted-ſkinned Melon.
——— *melo ſtriatus.* Late ſmall ſtriated Melon.
——— *melo tuberoſus.* Warted Cantaleupe.

* Linnæus defines an *Apple* to be a pulpy ſeed-veſſel, without a valve; and containing within it a membranous capſule, with ſeveral cells to receive the ſeeds.

S 4

Cucumis

Cucumis *melo turbinatus*. Top-shaped
Melon.

———— *melo virens*. Green rinded
Melon.

2 Cucumis chate. *Egyptian Melon*.

3 Cucumis fativus. *Common prickly Cucumber*.

———— *fativus albus*. White prickly
Cucumber.

———— *fativus longus*. Long prickly
Cucumber.

4 Cucumis flexuofus. *Green Turkey Cucumber*.

———— *flexuofus albus*. White Turkey Cucumber.

5 Cucurbita lagenaria. *Bottle Gourd*.
6 Cucurbita citrullus. *Water Melon*.
7 Cucurbita pepo. *Common Pompion*.

———— *pepo oblongus*. Long Pompion.
8 Cucurbita verrucofa. *Warted Gourd*.
9 Cucurbita melopepo. *The Squafh*, or
Melon Gourd.

10 Melothria pendula. *Small Creeping Cucumber*.

1 Cucumis melo. *Mufk Melon*. *Lin.
Sp. pl.* 1436.

Melo vulgaris. *Bauh. Pin.* 310.

What particular country the *Mufk Melon*
is a native of is not known, but it is now
cultivated in almoft every part of Europe.
The varieties mentioned in the lift are the
most

moſt diſtinguiſhed ones, but ſome of them are not worth the expence of raiſing. The ſmall *Portugal Melon* is a tolerable good one, and is the more to be eſteemed becauſe it comes early, and is a plentiful bearer.

The *Cantaleupe* is a middle-ſized fruit, of a roundiſh form, the outer coat is ſtudded with rough knobs, or protuberances like warts, the fleſh is generally of an orange colour, of a delicious flavour, and may be eaten in conſiderable quantities, without hurt to the ſtomach, which is not the caſe of moſt of the other ſorts. The Dutch are ſo fond of this that they pay little regard to any other, and by the way of pre-eminence, call it only *Cantaleupe,* not joining Melon to it. It takes its name from a place called *Cantaleupe,* about fourteen miles from *Rome,* where it is greatly cultivated, and where the Pope has a country-ſeat. But Miller ſays it was firſt brought thither from that part of *Armenia,* bordering on *Perſia,* in which place it is produced in ſuch plenty, that a horſe-load is ſometimes ſold for a French crown.

2 CUCUMIS chate. *Egyptian Melon. Lin. Sp. pl.* 1437.

Cucumis Ægyptius rotundifolius. *Bauh. Pin.* 310.

This is an annual, and grows ſpontaneouſly in Egypt. It hath long procumbent,

bent, obfolete angled ftalks, which put
forth clafpers, and are furnifhed with erect,
pellucid, white hairs. The leaves are almoft
round, and, like the ftalks, are covered with
a plufh of foft white hairs. The fruit alfo
is hairy, long, tapering, and the flefh almoft
of the fame confiftence as that of other
Melons. Miller reports that it is of an infipid tafte, and not worth cultivating; probably it may be fo here, for want of proper
management, or a natural foil and climate;
but in Egypt it is in fo much efteem, as to
have obtained the name of *Queen of Cucumbers.* The tafte is fweet, and a little watery. Haffelquift afferts, that the Grandees
and Europeans in Egypt, eat thefe as the
moft pleafant and refrefhing fruit they have,
and thofe from which they have the leaft to
apprehend; that they are the moft excellent of this tribe of any yet known, and that
the Nobles of Europe might wifh them at
their tables.

The plant is found in the fertile plains
round Cairo, after the inundation of the
Nile, and not in any other place in Egypt,
nor in any other foil.

3 Cucumis fativus. *Common Cucumber.*
Lin. Sp. pl. 1437.

Cucumis fativus vulgaris. *Bauh. Pin.*
310.

The *Common Cucumber* is another of thofe
plants

plants whofe native country is not known.
It is univerfally cultivated in all the four
quarters of the globe. The methods of eating the fruit here are too well known to
require any thing faid about them, but in
Egypt they have one perhaps peculiar to
themfelves: this is to fcoop out the chief
of the flefh, and fill the fhell with flefh and
aromatic herbs, and then boil it in the
manner of a pudding, which is faid to be
extremely palatable, and fatisfactory. In
fome parts of the Eaft they boil the fruit
whole, and eat them with falt and vinegar.
The feeds of Cucumbers, and thofe of the
Melon, are two of the greater cold feeds,
are deemed balfamic, cooling, and emollient, and are prefcribed amongft diuretics.

4 Cucumis flexuofus.   *Green Turkey
Cucumber.   Lin. Sp. pl.* 1437.
Cucumis oblongus.   *Bauh. Hift.* II.
p. 247.
This is fuppofed to be a native of India.
The ftalks and leaves are longer than thofe
of the former, and the fruit are fmooth,
and generally double the length of the
Common Cucumber. The variety, called the
White Turkey, is lefs watery than the green,
and therefore is more generally efteemed;
but the beft forts are counted unwhole-
fome, and by their coldnefs, apt to difpofe
the blood to putrid fermentations, and lay
6   the

the foundation of many of thofe malignant fevers, which often appear in autumn. To prevent thefe effects, therefore, they fhould always be eaten with plenty of falt, pepper, and vinegar.

5 CUCURBITA lagenaria. *Bottle Gourd.* *Lin. Sp. pl.* 1434.

Cucurbita oblonga, flore albo, folio molli. *Bauh. Pin.* 313.

The *Bottle Gourd* is a native of America, and is there much cultivated. This is the moft conftant fpecies of the genus, in regard to the form of its fruit. When the plant is in a foil that fuits it, the ftalks run to a prodigious length, and are covered with a fine, foft, hairy down. The leaves are large, heart-fhaped, toothed on their edges, with two glands each at their bafe, and woolly like the ftalks. The flowers are bell-fhaped, are large and white, have reflexed brims, and are fupported on long peduncles. The fruit is pear-fhaped, moftly a little bent inwards, and when ripe, the rind is woody, and of a pale yellow colour.

In both the Indies this plant is much cultivated, and the fruit fold in the markets for the table. In thefe parts they make a principal part of the food of the common people, for three or four months fucceffively. The inhabitants boil and eat them with vinegar. The large full grown fruit they

frequently

frequently fcoop, and filling the fhells with
meat and rice, boil them as a pudding.
Thefe fhells being hard and ligneous, ferve
them for funnels, and many other houfhold
utenfils.

6 Cucurbita citrullus.   *Water Melon.*
Lin. Sp. pl. 1435.
Anguria Citrullus dicta.   *Bauh. Pin:* 312.

The *Water Melon* is a native of the fouthern parts of Italy, and is not only much
cultivated there and other parts of Europe,
but alfo in Afia, Africa and America.   It is
an annual plant, and varies very much in
the fize, fhape, and colour of both its fruit
and the feeds; the latter are black in fome,
red in others, and the flefh yellow or red.
The leaves are cut and divided into many
parts, even almoft to the midrib.   The poor
people in Perfia, and the Levant, live almoft
entirely upon thefe, Mufk Melons, Cucumbers, and milk, during the hot months.
They are cooling, diuretic, and very wholefome, if ufed in moderation.   In Egypt,
fays Haffelquift, they furnifh the inhabitants
with meat, drink, and phyfic.   When the
fruit is perfectly ripe, they make a hole in
it, where the juice foon collecting, affords
them a hearty draught; and in burning
fevers, this liquor is mixed with rofe-water,
and a little fugar, and given the patient with
great fuccefs.   The unripe fruit are eaten

3   with

with bread, when in feafon, and by the
common people counted their beſt proviſion,
as they are obliged to put up with worſe
fare all the remaining part of the year. Notwithſtanding this, ſtrangers ſhould be cautious of making too free with them at firſt,
eſpecially in the heat of the day, as they are
apt to chill the blood too much, and thereby
occaſion cholics and violent fluxes.

   7 Cucurbita pepo. *Common Pompion.*
Lin. Sp. pl. 1435.
   Cucurbita major rotunda, flore luteo,
folio aſpero. *Bauh. Pin.* 213.
   The *Common Pompion* is cultivated all
over England, and the country people frequently raiſe it upon their dunghills, where
it often bears very good fruit. The leaves
are large, rough, and lobed, and the flowers
yellow. The fruit are roundiſh, ſmooth,
and yellow, and the ſeeds are ſwelled, or
puffed up at their margins.
   Many people eat this fruit, after they
have prepared it in the following manner:
they cut a piece from the ſide, and take out
the pulp, which they clear from the ſeeds,
and mixing it with ſliced apple, ſugar, and
ſpice, then fill the ſhell with the compoſition, and bake the whole in an oven.
When ſufficiently done it is brought to
table, where it furniſhes them with a hearty
meal.

meal. The native place of the plant is not
known.

8 CUCURBITA verrucofa. *Warted Gourd.*
Lin. Sp. pl. 1435.

This is an annual, and the plant is in fo
many refpects like the *pepo*, as hardly to be
diftinguifhed from it; but the fruit is
fmaller, the fhell more woody, and ftudded
with knobs or warts. Some people boil
thefe fruits, and efteem them delicate, but
for what good qualities I know not. The
Americans, however, cultivate them on purpofe for the table, and, when about half
grown, boil and eat them with their meat.
Where the plant grows naturally has not
yet been afcertained.

9 CUCURBITA melopepo. *The Squafh.*
Lin. Sp. pl. 1435.

Melopepo clypeiformis. *Bauh. Pin.* 312.

The *Squafh* is alfo an annual, has lobate
leaves like the former, but the ftalk is
moftly ftrong, bufhy, and erect. It puts
forth clafpers, although it does not climb,
nor is it procumbent. The fruit is knobby,
depreffed, or fhield-fhaped. The native
place of the plant is not known, but it is
much cultivated in North America, where
the inhabitants boil the fruit, when about
the fize of large Walnuts, and eat them as
the former.

10 MELOTHRIA

10 MELOTHRIA pendula. *Small Creeping Cucumber.* Lin. Sp. pl. 49.

This is an annual, a native of America, and the only plant at prefent known of the genus. It fends forth many trailing ftalks, which extend to a great length, and ftrike root at every joint; thefe are furnifhed with angular leaves, refembling thofe of the Melon, but they are not fo large. The flowers are of a pale fulphur colour, and each is compofed of a bell-fhaped, monophyllous calyx, having five teeth (the upper one of which often falls off) and a wheel-fhaped petal, fnipped at the edge into five obtufe fegments, with three conical filaments, tipped with twin, compreffed fummits, and inferted into the tube of the petal. The germen is an oblong-oval, and fupports a cylindrical ftyle, crowned with three oblong ftigmata, and becomes a fmooth, black, oval berry *, about the fize of a floe.

The inhabitants in the Weft Indies pickle thefe berries, and ufe them as we do Capers.

---

* This plant ought to have been placed in the Vth Chap. but as its general habit much refembles fome of the plants juft now defcribed, I judged it would be as well to fet it after them.

S E C T. II.

Apples of Trees.

1 ACHRAS ſapota. *Oval-fruited Sa-*
 pota.
2 Averrhoa carambola. *Goa Apple,* or
 Starry Plum.
3 Averrhoa bilimbi. *Bilimbi.*
4 Punica granatum. *Pomegranate-tree.*
5 Pyrus communis. *Pear-tree.*
6 Pyrus malus. *The Crab-tree.*
7 Pyrus cydonia. *Quince-tree.*

1 ACHRAS ſapota. *Oval-fruited Sapota.*
Lin. Sp. pl. 470.

Anona foliis laurinis glabris viridi-fuſcis,
fructu minore. *Sloane's Jam.* 206. *Hiſt.* II.

This tree is a native of South America,
and is commonly planted in their gardens
there. It riſes to about thirty feet high,
breaking into many branches, which form a
regular head, and are furniſhed with leaves,
ſhaped like thoſe of the Laurel, but are near
a foot long, two or three inches broad, and
of a browniſh-green colour. The flowers
are produced from the ſides of the branches,
ſtanding ſingly, and are of a cream colour.
Each has a permanent calyx, compoſed of

T   five

five oval, acute-pointed leaves, furrounding
five heart-fhaped petals, ending in an acute
point, and joined together at their bafe. In
the centre of thefe are five fhort awl-fhaped
ftamina, and one ftyle, longer than the petals, ending with an obtufe ftigma. The
germen is roundifh, and becomes an oval,
fucculent Apple, enclofing two or three
oval feeds. There is a variety of this tree,
bearing top-fhaped fruit, with fharp-pointed
feeds, and having a ruffet-coloured coat.
This laft is the cultivated fort.

The pulp of this fruit has a lufcious tafte,
refembling that of marmalade of Quinces,
whence it is called natural marmalade. The
ftones taken in emulfion are reckoned good
againft the gravel.

2 AVERRHOA carambola. *Starry Plum.*
Lin. Sp. pl. 613.
Mala goënfia, fructu octangulari pomi
vulgaris magnitudine. *Bauh. Pin.* 433.
This grows on the coaft of Malabar,
where it gets to the fize of a fmall Apple-
tree. It puts forth many branches from
the top, from which fhoot many flexile
twigs, furnifhed with oval, fharp-pointed,
dark-green leaves, of a rough bitterifh tafte.
The flowers come out at the joints of the
twigs, upon fhort peduncles; they have a
permanent, pentaphyllous calyx, furrounding five fpear-fhaped, blufh-coloured petals,
                              including

Including ten hair-like ftamina, tipped with roundifh fummits, and five fhort ftyles, crowned with fimple ftigmata. The germen is oblong, octangular, and becomes a yellowifh, eight-cornered fruit, about the fize of an hen's egg, containing many fmall angular feeds.

Thefe Apples have a pleafant acid tafte, are very cooling, and grateful to the ftomach.

3 AVERRHOA bilimbi. *Bilimbi. Lin. Sp. pl.* 613.

This grows in the fame parts of India as the former, and differs little from it except in the angles of the fruit; they being in this fpecies obtufe, and in the *carambola* acute; a difference not attended to by travellers, which occafioned their confounding them as one.

4 PUNICA granatum. *Pomegranate-tree. Lin. Sp. pl.* 676.

Malus punica fylveftris. *Bauh. Pin.* 438.

This is a native of Spain, Portugal, and Italy. It hath a woody ftem, which rifes fixteen or eighteen feet high, fending out many branches, garnifhed with fhining-green, fpear-fhaped leaves, ftanding oppofite. The flowers proceed from the extremities of the branches, fome ftanding fingly, and others three or four together,

regularly

regularly expanding in their turns, by which there is a fucceffion of flowers for a confi-derable time. The calyx confifts of a bell-fhaped, red, flefhy leaf, cut at the brim into five fharp fegments, and includes five roundifh fcarlet petals, inferted into the bottom of the calyx, as are the ftamina, which are many in number, very flender, and furround one ftyle, longer than them-felves. The germen is roundifh, and fwells to a large round fruit, having a hard reddifh rind, crowned with the remains of the ca-lyx, and contains many roundifh, fucculent feeds.

The flefh of thefe fruits is of a yellowifh colour, and a vinous flavour, but it is fub-ject to generate wind, and caufe pains in the ftomach and bowels. They fhould always be eaten cautioufly, left they throw the blood into a ftate of putrefaction.

There are feveral varieties of this tree now cultivated in gardens, and two or three with double flowers; the calyces of the latter are the Balauftines of the fhops, and are of an aftringent nature.

5 PYRUS communis. *Pear-tree.* *Lin. Sp. pl.* 686.

Pyrus fylveftris. *Bauh. Pin.* 439.

This grows wild in the woods and hedges of England. The generic characters are: the flower hath a permanent calyx of one
concave

concave leaf, divided into five fegments at the margin, and five concave petals, inferted into it. The ftamina are about twenty in number, are awl-fhaped, fhorter than the petals, and are inferted into the calyx. The germen is round, feated under the flower, and fupports five erect ftyles, crowned with fingle ftigmata. The fruit is large, flefhy, hath five membranaceous cells, each containing one fmooth, oblong, pointed feed.

Neither *Pears* nor *Apples* in their wild ftate are of much value, but art and induftry have obtained many varieties from them, which can hardly be excelled by any fruits in the world. Nor do any add more to the œconomy of human life than thefe; for befide the pleafure and refrefhment they afford when eaten raw, they furnifh excellent pies, tarts, and other devices, and ornament the table with the wholefome and cooling liquors of *Cider* and *Perry*. In fetting down the varieties of the *Pear*, I fhall reject fuch as are of an ordinary quality, and divide the reft into three Claffes: the firft comprehending fuch as are adapted for the table; the fecond fuch as are well enough qualified for this purpofe, but degenerate when grafted on Quince-ftocks; and the laft, thofe that are proper for baking.

T 3   C L A S S

C L A S S I.

1 *Petit Muſcat*, or *Su-preme*.
2 *Little Baſtard Muſk*.
3 *Early Ruſſet*.
4 The *Magdalen*.
5 *Great Blanquette*.
6 *Muſk Blanquette*.
7 *Long-ſtalked Blanquette*.
8 *Red Orange*.
9 *Auguſt Muſkat*.
10 *Summer Boncretien*.
11 *Swan's Egg*.
12 *Princes' Pear*.
13 *Roſewater*.
14 The *Red Rutter*.
15 *Summer Bergamot*.
16 *Autumn Bergamot*.
17 The *Rouſſeline*.
18 The *Royal Muſcat*.
19 The *Jargonelle*.
20 The *Melting Muſk*.
21 *Red Bergamot*.
22 *Swiſs Bergamot*.
23 *Late Bergamot*.
24 *Fig Pear*.
25 *German Muſcat*.
26 *Dutch Bergamot*.
27 *St. Martial*.
28 *St. Germain*.
29 *Chaumontelle Wilding*.
30 The *Autumn Beauty*.
31 *Good Lewis*.
32 *Grey Dean*.
33 *Winter Thorne*.
34 The *Royal Winter*.
35 The *Marchioneſs*.
36 *Winter Orange*.
37 The *Donville*.
38 *Winter Ruſſelet*.
39 *Beautiful Winter*.
40 The *Saraſin*.

The *Little Muſk*, or *Supreme Pear*, is rather round than long, and is generally produced in cluſters. The ſtalk is ſhort, the ſkin yellow, the juice a little muſky, and is beſt flavoured when not too ripe, which is early in July *.

The *Little Baſtard Muſk* is ſhaped like the *Su-*

---

* The ſummer 1782 being a very unkind one for ripening fruit, more kind ſeaſons may perfect ſome of them a fortnight or more ſooner than here mentioned.

preme,

preme, but is fmaller. It is feldom produced in clufters, and the fide next the fun has a few ftreaks of red. It ripens much at the fame time with the former, and it is more valuable for coming early, than for its extraordinary qualities.

The *Early Ruffet* is a fmall top-fheaped Pear, with a yellow fkin, dafhed with red and grey on the funny fide; the flefh is yellowifh, half-breaking, a little ftony next the kernels, and has a perfumed, fugary juice.

The *Magdalen* is a middling-fized fruit, rather long, of a greenifh-yellow when ripe; the flefh is white, melting, the juice perfumed, fweet, and mixed with a pleafant acid.

The *Great Blanquette*, or *Bagpipe of Anjou*, is a pretty large Pear, approaching to round. The fkin is fmooth, of a pale green colour, and full of a rich-flavoured juice. The ftalk is fhort, thick, and fpotted, and the leaf is like that of the Jargonelle. It ripens early in Auguft.

The *Mufk Blanquette* is a fmall fruit, much lefs than the former, and more pinched in at the ftalk, which is about the fame length with the other, but flenderer. The fkin is foft, of a pale green, the flefh tender, and full of a rich mufky juice. It ripens rather later than the Blanquette.

The *Long-ftalked Blanquette* is fhaped like the *Mufk*, but it is more hollowed at the crown, and has a larger eye. It is plumpifh towards the ftalk, and a little crooked. The fkin is fmooth, of a greenifh-white, fometimes has a ruffet tinge on the funny fide. The flefh is white, partly breaking, and plentifully ftored with a vinous, fugary, perfumed juice. It ripens with the former.

The *Red Orange* is a middling-fized round Pear, much the fhape of a Bergamot; of a greenifh colour,

lour,

lour, except next the sun, where it is often purple, or red. The stalk is short, the eye very hollow, the flesh melting, and the juice sugary and musky.

The *August Muscat*, or the *Royal Pear*, is very much shaped like a Bergamot. The stalk is long, straight, a little spotted, and the eye a little hollowed. The skin is smooth, of a whitish yellow colour, the flesh breaking, and the juice very sugary and much perfumed. It ripens at the end of August, and is esteemed one of the best Pears the summer produces.

The *Summer Boncretien*, or *Good Christian*, is a large oblong Pear, with a thin, smooth, whitish green skin, except on the sunny side, where it is of a good red. The flesh is between breaking and tender, and is stored with a rich juice, of a high perfumed flavour. It ripens early in September.

The *Swan's Egg* has its name from its shape. The skin is of a green-yellow, and striped with a russet-red and green on the sun-side. The flesh is firm, a little melting, the juice sugary, slightly musky, but of an agreeable flavour.

The *Princes' Pear* is a small roundish, yellowish fruit, except next the sun, where it is of a bright red. The flesh is between melting and breaking, and the juice highly flavoured. It ripens in September, and is the more valuable because it is a good bearer.

The *Rosewater* is a large round Pear, rather flattish, hath a very short stalk, at the insertion of which it is hollowed like an apple. The skin is rough, of a brown colour, the flesh breaking, the juice very sweet, and it becomes ripe in September.

The

The *Red Butter, Grey Butter,* or *Green Butter,* is of different colours, according to the ftock it hath been grafted upon. When propagated upon a free ftock it is brown. As to its general fhape, it is large and long. The flefh is very melting, full of a rich fugary juice, and it becomes ripe about the middle of October.

The *Summer Bergamot,* or *Hemden's Bergamot,* is a pretty large, flattifh Pear, of a greenifh-yellow colour, and hollowed at both ends like an apple. The flefh is melting, the juice highly perfumed, and it ripens a little before the former.

The *Autumn Bergamot* is a fmaller fruit than the former, but much of the fame fhape. The fkin is of a faint red on the funny fide, but of a yellowifh-green on the other; the flefh is melting, and when ripe, which is in the beginning of October, the juice is highly perfumed.

The *Rouffeline,* or *Long-ftalked Autumn Mufcat,* is a fmallifh Pear, having a fmooth fkin, of a greenifh-yellow colour, except on the funny fide, where it is red, with fome fpots of grey. The ftalk is long, the flefh tender, delicate, and very fweet, with an agreeable perfume. It ripens towards the end of October.

The *Royal Mufcat* is a fmall top-fhaped fruit, with a roughifh grey fkin, inclining to brown next the fun. The flefh is white and coarfifh, but the juice is fweet, mufky, and tolerably agreeable.

The *Jargonelle* is a long top-fhaped fruit, of a fine red colour next the fun, but very yellow on the fhady fide. The flefh is white, half breaking, tolerable fine, and the juice a little mufky.

The *Melting Mufk* is alfo a long top-fhaped Pear, of a middling fize. The fkin is even,   —

fmooth,

smooth, of a grafs-green round the apex, but of a yellowifh one near the ftalk. The flefh is melting, the juice high flavoured, and very mufky.

The *Red Bergamot* is rather a fmallifh Pear, top-fhaped, and flatted; next the fun it is of a yellow-red colour; the flefh is melting, the juice high flavoured, and very perfumed.

The *Swifs Bergamot* is a roundifh Pear, with a tough, greenifh-coloured fkin, ftriped with red. The flefh is melting and full of juice, but is not fo richly perfumed as the former. It ripens the beginning of October.

The *Late Bergamot*, *Colmar*, or *Manna Pear*, is fomewhat like a Boncretien, but the head is flat, the eye large and deeply hollowed. It is thickeft in the middle, floping toward the ftalk, which is fhort, thick and a little bent. The fkin is green, with a few yellow fpots, and fometimes it is a little coloured next the fun; the flefh is tender, and the juice greatly fugared.

The *Fig Pear* is a middling fized fruit, of a long top-fhape. The fkin is rather fmooth, of a brownifh-green when ripe, and the flefh white and melting. The juice is fweet, fugary, and heightened with a pleafant fharpnefs. Ripe the beginning of October.

The *German Mufcat* is rather a long top-fhaped Pear, much of the form of the Royal Winter, but more contracted near the eye; the fkin too is of a more ruffet colour, and red on the funny fide. The flefh is melting, buttery, and a little mufky.

The *Dutch Bergamot* is fhaped like the Common Bergamot, but it is a larger fruit. The juice is highly flavoured, the fkin greenifh, and the flefh half buttery and tender.

The *St. Martial*, or the *Angelick Pear*, is oblong,

5   much

much the shape of the Boncretien, but it is not so large, and a little flatter at the crown. The stalk is very long, the skin smooth and yellowish, except next the sun, where it is generally purplish. The flesh is melting, the juice very rich, and a little perfumed. It is a late fruit, and counted one of the best yet produced.

The *St. Germain* is a large, long Pear, of a yellowish-green colour, and melting. In dry seasons it abounds with a sweet agreeable juice, and is a very good fruit, but in moist ones, or on damp soils, it is roughish and austere. It is in eating for about two months after Christmas.

The *Chaumontelle Wilding* is rather a large Pear, and flatted at the crown. The skin is roughish, of a pale green colour, except on the sunny side, where it is purplish. The flesh is melting, the juice very rich, and a little perfumed. This is esteemed an excellent fruit, and is in eating from November to January.

The *Autumn Beauty* is a pyramidal-shaped Pear, with a tolerable smooth skin, of a fine deep red next the sun, speckled with grey. The shady side is partly red, but not so deep, and partly yellow, speckled with fawn colour. The flesh is white, breaking, sometimes half melting, the juice copious, and of a high flavour.

The *Good Lewis* is nearly of the shape of the St. Germain, but is not quite so pointed. The stalk is very short, a little bent, the skin very smooth, and the eye small. When ripe it is of a whitish-green colour, and if it grow upon a dry soil, the flesh will be very tender, and full of a rich sweet juice. It is in eating in December.

The *Grey Dean* is a middling-sized, roundish Pear; the skin smooth, of a greenish-grey colour, the

the flefh buttery, melting, and not fubject to be
woolly like the Yellow Dean. The juice is very
fugary, and of a tolerable good flavour. It ripens
in November.

The *Winter Thorn* is rather a large Pear, of a
pyramidal figure, the fkin fmooth, of a whitifh-
green at firft, but of a pale yellow when ripe.
The ftalk is fhort and flender ; the flefh melting
and buttery, the juice very fweet, and, if the feafon prove dry, highly perfumed. It ripens at the
end of December.

The *Royal Winter* is a large top-fhaped fruit,
with a fine fmooth, beautiful red fkin on the
funny fide, and when ripe, yellow on the other.
It is often fpeckled with brown fpots upon the
red, and fawn-coloured upon the yellow. The
flefh is inclining to yellow, is very fine, half
buttery, melting, and on dry foils the juice is
very fugary. It ripens in December.

The *Marchionefs* is a large pyramidal Pear, of a
green colour at firft, with dots of a deeper green ;
but when ripe becomes yellow, and frequently
with a flight tinge of red. The flefh is melting,
buttery, the juice fweet, fugary, and fometimes a
little mufky. Ripe the beginning of December.

The *Winter Orange* is a middle-fized fruit, of
the fhape of an Orange. The fkin is ftudded
with fmall knobs, and is of a pale brown-green
when ripe, with fome little dots of a browner
green. The flefh is white, fine, breaking, and
the juice mufky and agreeable. Ripens in February.

The *Donville* is a middle-fized Pear, fharpifh at
both ends, the fkin fmooth and fhining, of a deep
lemon colour, and fcattered with fawn-coloured
fpots on the fhady fide, but of a bright red, fpeckled
with fmall grey dots on the other. The flefh is
                                             inclining

inclining to yellow, it is breaking, and the juice is highly flavoured, with a little fharpnefs. Ripens in February.

The *Winter Ruffelet* is a fmall top-fhaped Pear, with the fkin partly greenifh and partly reddifh. The flefh is half breaking, copioufly ftored with juice, which is of a tolerable high flavour. Ripens at the end of February.

The *Beautiful Winter* is a pretty large fruit, and nearly round. The fkin is fmooth, and yellow on the fhady fide, fpeckled with fawn-colour; but on the funny fide it is of a beautiful red, fpeckled with bright grey. The flefh is tender, the juice copious, and of a pleafant fweetnefs. It ripens in February.

The *Sarafin* is the moft valuable of all the Pears for duration, as it will keep found both upon, or off the tree for twelve months. It is of a middle fize, about a third part longer than broad, the fhady fide of a pale yellow when ripe, but the funny fide of a brownifh red, fpeckled with grey. The flefh is white, almoft buttery, the juice fugary, highly flavoured, and a little perfumed.

CLASS II.

Pears which degenerate when grafted on Quince-ftocks.

1 *Meffire John.*	6 The *Little Lard.*
2 The *Green Sugar.*	7 The *Ronville.*
3 The *Dauphine.*	8 The *Gate.*
4 The *Dry Martin.*	9 The *Eafter Bergamot.*
5 The *Large-ftalked.*	10 The *Winter Boncretien.*

The *White and Grey Meffire John* are deemed one and the fame fruit, the difference of their colour

being

being occafioned by the foils they may grow in, or the ftocks they may be grafted upon. It is a large roundifh Pear, moftly having a brown, rough fkin. If grafted on a free-ftock, and planted in a moift foil, the flefh will be breaking, and copioufly ftored with a rich fugary juice; but on a Quince-ftock it will be harfh and ftony. Ripe in October.

The *Green Sugar* is fhaped like the Winter Thorn, defcribed in the former Clafs, but is fmaller. The fkin is very fmooth, green, and the flefh buttery, fugared, and of a good flavour; but if grafted on a Quince-ftock, it will be ftony. It ripens at the beginning of November.

The *Dauphine*, or *Lanfac*, is a top-fhaped Pear, about the fize of a Bergamot, flatted near the head, but a little lengthened near the tail. It is fmooth, of a yellowifh-green colour on the outfide, yellow within, the flefh tender and melting, the juice fugared, and flightly perfumed. The eye is very large, and the ftalk long and ftraight. It ripens in November, and if planted in a good foil, and grafted on a free-ftock, it is one of the beft table Pears then in feafon.

The *Dry Martin*, or *Champagne*, is much like the Ruffelet both in fhape and colour, but it is rather more oblong. The flefh is fine and breaking, and the juice fugared, with a flight perfume, and if grafted on a free-ftock, is an excellent Pear. It comes in eating at the end of November.

The *Large-ftalked* is a yellow, roundifh Pear, with a very thick ftalk, whence it had its name. The flefh is dry, breaking, and has a mufky flavour; it is much improved by being planted in a moift foil, and grafted on a free-ftock. It comes in eating with the former.

The

The *Little Lard*, or *Anjou Ruffet*, alfo the *Winter's Wonder*, is a middle-fized fruit, but is apt to vary in fhape, it being fometimes nearly oval, and at others refembles a Bergamot. The fkin is a little rough, greenifh at firft, but turns yellowifh when ripe, and is fprinkled with little knobs. The ftalk is long and flender, the eye large, and deeply hollowed; the flefh fine, buttery, and melting, the juice fugary, mufky, and of an agreeable flavour, but is much hurt when grafted on a Quince-ftock. It ripens at the beginning of November.

The *Ronville*, or *Lord Martin*, is about the fize of a large Ruffelet, but the middle of the Pear is moftly fwelled more on one fide than on the other, and the eye is hollowed a little. The fkin is foft, very fmooth, of a lively red next the fun, but when ripe, of a yellow on the other. The flefh is breaking, full of juice, which is very fweet, and a little perfumed. On a Quince-ftock it is apt to be ftony.

The *Gate* is a round Pear, and has a fweet, fugary juice, a little perfumed, if grafted on a free-ftock, and planted in a rich foil; but in a dry foil, and upon a Quince-ftock, it is good for nothing.

The *Eafter Bergamot* is a large Pear, and nearly round, except towards the ftalk, where it lengthens a little. The eye is flat, the fkin at firft green, but turns yellow when ripe, with fmall brown dots, and a tinge of red on the funny fide. The flefh is fine, inclining to yellow, and is buttery and melting. If grafted on a free-ftock the juice is very fweet, fugary, and high flavoured. It ripens in January.

The *Winter Boncretien* is a very large Pear, of a
pyramidal

pyramidal form, flat at the top, the skin very fine, of a bright yellow colour, inclining to green, but of a soft flesh-red on the sunny side. If planted in a good soil, and grafted on a free-stock, the flesh will be fine, tender, full of a sweet, sugary juice, of a perfumed, vinous flavour. Ripens in January.

C L A S S   III.

Pears proper for Stewing and Baking.

1 Le *Besidéri*, or *Heri*.	6 The *Catillac*.
2 The *Spanish Boncretien*.	7 The *Double flower-ing*.
3 The *Pound*, or *Lovely Pear*.	8 The *Burnt Cat*.
4 The *Winter Citron*.	9 The *Pope's Pear*.
5 The *Golden End of Winter*.	10 The *Union*.

The *Le Besidéri* is a middling-sized round Pear, of a pale green colour, inclining to yellow. The stalk is very long and slender, and the flesh dry. It ripens near the end of November.

The *Spanish Boncretien* is a large pyramidal fruit, of a pale yellow colour on the shady side, but of a fine lively red on the other. The skin is smooth, and all over speckled with small brown dots. The flesh is white, mixed with greenish spots, and it is either tender, hard, dry, or juicy, according to the soil, season, or stock it may be grafted on. Ripe at the end of November, or beginning of December.

The *Pound*, or *Lovely Pear*, also *Parkinson Warden*, is a large fruit, which commonly weighs a pound or more. The skin is rough, of a dull red

next

next the fun, but fomewhat paler on the other fide. The ftalk is very fhort, and the eye much hollowed. Comes in feafon in December.

The *Winter Citron*, or *Mufk Orange*, is a tolerable large Pear, nearly of the fhape and colour of an Orange. It is an ordinary Pear for the table, but will bake well, and is in feafon with the former.

The *Golden End of Winter* is a very large fruit, almoft of a globular form. The ftalk is fhort, the fkin yellow, fpotted with red, the flefh dry, and very apt to be ftony. Comes in feafon in January.

The *Catillac* is a large Pear, and nearly of the fhape of a Quince. The fkin is generally yellow, but turns to a deep red on the funny fide. The flefh is hard, the juice auftere, yet it bakes well. Comes into ufe in January.

The *Double-flowering Pear* is a thick, fhort fruit, with a long, ftraight ftalk. The fkin is very fmooth, of a yellowifh colour, except on the funny fide, where it is moftly red or purple. It is a moft excellent Pear for baking, and comes in feafon in February. The flower having two ranges of petals obtained it the name it goes by.

The *Burnt Cat* is rather a fmall Pear, of an oblong form. The fkin is fmooth and fhining, reddifh next the fun, but of a fort of lemon colour on the other. The flefh is tender, but dryifh, and acquires in baking a beautiful red. It ripens in February.

The *Pope's Pear* is of a middling fize, and common fhape. The fkin is roughifh, yellow, or inclining to a cinnamon colour. The flefh tender, white, and moftly without ftones.

The *Union* is a large, long Pear, of a reddifh

U   colour

colour next the fun, but of a deep green on the other fide. It comes in feafon in January, is a good baking Pear, and a plentiful bearer.

6 PYRUS malus. *The Crab-tree.* *Lin. Sp. pl.* 686.

Pyrus foliis ferratis, pomis bafi concavis. *Hort. Cliff.* 189.

The *Crab-tree* is common in every part of England, and is the parent of all the *Apple-trees* at prefent cultivated. Its varieties are fo exceedingly numerous, that it is impoffible for any one clearly to afcertain them; for even in its wild ftate, almoft every different foil and fituation the feeds may chance to vegetate in, produce fome fmall variation in the form, colour, or flavour of the fruit. It is remarkable that the *Crab*, or *Apple-tree*, though it exactly agrees in the generic characters of the fructification, with thofe of the *Pear* and *Quince*, yet it will not take when grafted upon either of them, nor they upon the *Apple*; which feem to indicate, that this genus is not a natural one *, and that nature has placed fome boundary between the latter, and the two former, but

* This was a main argument with Miller for fplitting the genus, and it was conftantly contradicted by his own experience as a gardener; for he acknowledges the *Peach* to be a diftinct genus from the *Plum*, and yet it is a common practice in the nurfery to bud the former either upon the latter or upon an *Apricot*, and they are found to take very well.

fuch

such as is beyond our penetration to dif-
cover. Linnæus certainly, therefore, did
right in placing them all under one genus,
and not separate them, as Miller and others
have done; as in any fyftematical arrange-
ment, we muft always be governed by what
is plain and obvious in the ftructure of the
plants, otherwife the defign will be rendered
abortive.

In fetting down the varieties of the cultivated *Apple*, I fhall defcribe only fome of
the moft valuable ones, and divide them into
two Claffes: the firft to contain fuch as are
immediately adapted to the table, in order
to be eaten raw; and the fecond to confift
of thofe proper for boiling, baking, &c.

C L A S S   I.

1 The *Summer Calville.*
2 The *Anife.*
3 The *Common Codlin.*
4 The *Margaret.*
5 The *Summer Pearmain.*
6 *Loan's Pearmain.*
7 The *Quince Apple.*
8 The *Ruffet Rennet.*
9 The *French Rennet.*
10 The *Rennet Grife.*
11 The *Red Rennet.*
12 The *White Calville.*
13 The *Red Calville.*
14 The *Aromatic Pippin.*
15 The *Golden Pippin.*
16 The *Violet Apple.*
17 The *Hollow Crown'd Pippin.*
18 The *Winter Rambour.*
19 The *Great Faros.*
20 The *Nonpareil.*

The *Summer Calville* is a middling-fized Apple,
of a longifh form, and the fkin is ftreaked with
red and white. The flefh is light and dry, of no

U 2   extraordinary

extraordinary flavour, but the fruit is efteemed for coming early.

The *Anife Apple* is a middling-fized fruit, of a greyifh colour, and rather longer than a Golden Pippin. The flefh is tender, and hath a fpicy flavour like Anife-feed or Fennel.

The *Common Codlin* is a large, early, good-flavoured Apple, and is too well known to require any defcription.

The *Margaret* is a middling-fized fruit, fhorter than the Codlin, and the fkin on the funny fide is of a faint red, the other fide of a pale green. The flefh is firm, and of a pleafant flavour, but foon decays.

The *Summer Pearmain* is an oblong Apple, and is ftriped with red on the funny fide. The flefh is tender, but it foon becomes mealy.

Lean's Pearmain is a middle-fized Apple, of a beautiful red on the funny fide, and is ftriped with red on the other. The flefh has a vinous, quick flavour, but it foon grows mealy.

The *Quince Apple* has its name from its fhape, which is like that of a Quince. It is about the fize of a Golden Pippin, but of a longer form, efpecially near the ftalk. It is of a ruffet colour on the funny fide, and inclining to a yellow on the other. The flavour is very agreeable.

The *Ruffet Rennet* is a fmall fruit. Its name fpeaks its colour. It will keep a long time, and the flefh has a high flavour.

The *French Rennet* is a large, roundifh, yellowifh-green Apple, dotted with fmall grey fpots. The juice is fugary, and of a good flavour. This is an excellent fruit for keeping.

The *Rennet Grife* is a middle-fized Apple, and is fhaped like the Golden Rennet; it is of a deep

grey

grey colour on the funny fide, but mixed with yellow on the other. The flefh is very juicy, and of a quick flavour.

The *Red Rennet* is fomewhat rounder than the former, and of a beautiful red colour, on a whitifh ground. The flefh is firm, and the juice fugary. It feems to be only a variety of the French Rennet.

The *White Calville* is a large, white, fquarifh Apple. The flefh has a high flavour, without any acid. It will keep a long time, which makes it much efteemed.

The *Red Calville* is a large, red fruit, and longer than round. The flefh of this is fometimes reddifh, and has a fine vinous flavour.

The *Aromatic Pippin* is near the fize of the Nonpareil, but a little longer. The fide next the fun is of a bright ruffet colour. The flefh is tender, and hath an aromatic flavour.

The *Golden Pippin* is a middle-fized fruit, of a yellow-gold colour, and is rather longer than round. It is dotted with fmall red fpots. Its juice is fugary, and very high-flavoured.

The *Violet Apple* is a pretty large fruit, of a greenifh white, ftriped with a deep red on the funny fide. The flefh is white, very fine, and the juice fugary, with fome faint flavour of a violet.

The *Hollow-crowned Pippin* is a middling-fized Apple, and very hollow at the top, whence its name.

The *Winter Rambour* is a very large fruit, and nearly round. It is quite green, and the juice has a fharp acid tafte.

The *Great Faros* is a large, flattifh Apple, ftreaked with red. The flefh is breaking, and plentifully ftored with juice.

The *Nonpareil* is a fmallifh fized fruit, rather

conical,

conical, of a ruffet-green colour, a little inclining
to red on the funny fide. The flefh has a fine
flavour, and is much efteemed.

C L A S S　　II.

Apples proper for boiling, baking, &c.

1　The *Summer Rambour*.　　6 The *Holland Pippin*.
2　The *Kentifh Fill-Bafket*.　7 The *Embroidered Apple*.
3　The *Golden Rennet*.　　　8 The *Royal Ruffet*.
4　The *Hertfordfhire Pear-*　9 *Wheeler's Ruffet*.
　　main.　　　　　　　　　10 *Pile's Ruffet*.
5　The *Kentifh Pippin*.

The *Summer Rambour* is a very large fruit, and
rather flatter than the Winter Rambour. The
fkin is white, with fome few ftreaks of red. It
comes early, and is an excellent Apple for ftewing.
The *Kentifh Fill-Bafket* is a large fort of Codlin,
but is longer than the Common Codlin. This is
a good baking Apple.
The *Golden Rennet* is proper either for eating
raw, or baking.
The *Hertfordfhire*, or *Winter Pearmain*, is a tolerable fized fruit, rather longer than round. It is
of a fine red on the funny fide, and ftriped with
the fame colour on the fhady one. The flefh is
juicy, and it ftews well.
The *Kentifh Pippin* is a large, oblong Apple, of
a pale green colour. The flefh is juicy and
breaking, of a quick acid flavour, and it boils
well.
The *Holland Pippin* is both a larger and longer
Apple than the former, and the fkin is of a darker
　　　　　　　　　　　　　　　　　　　　　green

green colour. It is firm and juicy, and boils well.

The *Embroidered Apple* is a largish fruit, and somewhat resembles the Winter Pearmain, but the stripes of red are broader. It is used as a kitchen Apple.

The *Royal Russet*, or *Leather Coat*, is a large, oblong Apple, with a deep russet-coloured skin. This is an excellent fruit for boiling, and a good one to eat raw.

Wheeler's Russet is a flat, middling-sized Apple. The side next the sun is of a pale russet-colour, the other is inclining to yellow. The juice has a very quick acid flavour, and it boils well.

Pile's Russet is of an oval figure, and is a smaller Apple than the former. The skin is of a russet-colour on the sunny side, and of a dark green on the other. The flesh has a quick acid taste, and it is a good fruit for baking.

There is a large number of valuable *Apples* yet remaining, but their appellations are so various in different places, that it is impossible to describe them by any certain general names. Those commonly used for the making Cyder are the following :

1 The *Red Streak*.
2 The *Devonshire Royal Wilding*.
3 The *Whitsour*.
4 The *Hertfordshire Underleaf*.
5 The *John Apple*.
6 The *Everlasting Hanger*.
7 The *Gennet Moyle*.
8 The *Cat's Head*.

7 Pyrus cydonia. *The Quince-tree. Lin. Sp. pl.* 687.

Malus

Malus cotonea fylveftris. *Bauh. Pin.* 434.

The *Quince-tree* grows naturally on the banks of the river Danube, in Hungary. It is a rather fmaller tree than the Crab. The leaves are nearly of the fame fhape, but have more prominent ribs, and are whiter on their under fide. The flowers come out fingly, and the calyx is ferrated, fpreading, and of the length of the petals. The fruit is very well known. The varieties of it are, the *Pear* and *Portugal Quince.* The laft is deemed the beft, and is the fort now moft generally cultivated. The flefh of this is lefs auftere than the other, of the fineft purple colour when ftewed, and it makes the moft agreeable and beft flavoured Marmalade.

Quinces are very aftringent; employed medically they ftrengthen the ftomach, and ftop fluxes of the bowels. A fyrup is frequently made of the juice, and prefcribed for thefe purpofes. The bruifed feeds impart a very ftrong mucilage to any watery liquor, which makes an excellent gargarifm for fore mouths. An ounce will render three pints of water as ropy as the whites of Eggs.

C H A P.

C H A P.   VIII.

L E G U M I N O U S * P L A N T S.

S E C T.   I.

Pods and Seeds of Herbaceous Plants.

1 ARRACHIS hypogæa. *American Ground Nut.*
2 Cicer arietinum. *The Chich Pea,* or *Gravances.*
3 Dolichos foja. *Eaſt India Kidney Bean.*
4 Ervum lens. *Lentil.*
5 Lotus edulis. *Incurved-podded Bird's-foot Trefoil.*
6 Lotus tetragonolobus. *Square Podded Crimſon Pea.*
7 Lupinus albus. *White-flowering Lupine.*
8 Phaſeolus vulgaris. *Common Kidney Bean.*
——— *coccineus.* Scarlet-flowering Kidney Bean.
——— *albus.* White-flowering Kidney Bean.
9 Piſum ſativum. *Common Garden Pea.*

* A Legumen is a pod with two valves, incloſing a number of ſeeds that are faſtened along one ſuture only.

Piſum

Pifum umbellatum. Rofe, or Crown Pea.

—— *quadratum.* Angular-ftalked Pea.

10 Pifum Americanum. *Cape,* or *Lord An-fon's Pea.*

11 Pifum maritimum. *Sea Pea.*

12 Vicia faba. *Common Garden Bean.*

—— *minor.* The Horfe Bean.

1 ARRACHIS hypogæa. *American Ground Nut. Lin. Sp. pl.* 1040.

This is an annual plant, and a native of Brafil and Peru. The ftalks are long, trail upon the ground, and are furnifhed with winged leaves, compofed of four hairy lobes each. The flowers are produced fingly on long peduncles; they are yellow, of the pea kind, and each contains ten awl-fhaped ftamina, nine of which are tyed together, and the upper one ftands off. In the centre is an awl-fhaped ftyle, crowned with a fimple ftigma. The germen is oblong, and becomes an oval-oblong pod, containing two or three oblong blunt feeds.

This plant is cultivated in all the American Settlements for the feeds, which make a confiderable part of the food of the flaves. The manner of perfecting them is very fingular, for as the flowers fall off, the young pods are forced into the ground by a natural motion of the ftalks, and there they are entirely buried, and not to be difcovered
without

without digging for them, whence they have taken the name of *Ground Nuts*.

2 CICER arietinum. *Chich Pea. Lin. Sp. pl.* 1040.

Cicer fativum. *Bauh. Pin.* 347.

The *Chich Pea* grows naturally among the corn in Spain and Italy, and it is much cultivated in thefe places for the table. It is an annual, fending up feveral hairy ftalks, near two feet high, which are fet with pinnated leaves, compofed of eight or nine pair of oval, ferrated pinnæ, with an odd one at the end. The flowers are fmall and whitifh, are of the pea kind, moftly but one on a peduncle, have ten ftamina each, nine of which are joined together, and the tenth ftands off. The germen is oval, and becomes a turgid, hairy, rhomboidal pod, containing two roundifh feeds, of the fize of common peas, each having a protuberance on the fide.

Though thefe Peas are common at table in Spain and Italy, they would badly fuit an Englifh ftomach, being far from delicate, but are ftrong, flatulent, and hard of digeftion. There are two varieties of this plant, one with red, and the other with black feeds. It is much cultivated in Barbary, by the name of *Gravances*, and is counted one of their beft forts of pulfe.

3 DOLICHOS

3 Dolichos foja. *Indian Kidney Bean.*
Lin. Sp. pl. 1023.

This is a perennial, and a native of India. It fends up an erect, flender, hairy ftalk, to the height of about four feet, furnifhed with leaves much like,thofe of the Common Kidney Bean, but more hairy underneath. The flowers are produced in erect racemi, at the bofoms of the leaves; they are of the pea-kind, of a bluifh white colour, and are fucceeded by pendulous, hairy pods, refembling thofe of the Yellow Lupine, each containing three or four oval, white feeds, a little larger than peas.

This plant is much cultivated in Japan, where it is called *Daidfu,* and where the pods fupply their kitchens for various purpofes; but the two principal are with a fort of butter, termed *Mifo,* and a pickle, called *Sooju* or *Soy.*

The *Mifo* is made by boiling a certain quantity of the beans for a confiderable time in water, till they become very foft, when they are repeatedly brayed with a large quantity of falt, till all is incorporated. To this mafs they add a certain preparation of Rice, named *Koos,* and having well blended the whole together, it is put into a wooden veffel, where in about two months it becomes fit for ufe, and ferves the purpofes of butter. The manner of preparing the *Koos* is a kind of fecret bufinefs, and is

2　　　　　　　　　　　　　in

in the hands of fome certain people only, who fell the *Koos* about the ftreets, to thofe who make *Mifo*.

In order to prepare *Sooju* they take equal quantities of beans, wheat, or barley-meal, and boil them to a pulp, with common falt. As foon as this mixture is properly incorporated, it is kept in a warm place for twenty-four hours to ferment; after which the mafs is put into a pot, covered with falt, and a quantity of water poured over the whole. This is fuffered to ftand for two or three months, they never failing to ftir it well at leaft once a day, if twice or thrice it will be the better; then the liquor is filtered from the mafs, and preferved in wooden veffels, to be ufed as occafions require. This liquor is excellent for pickling any thing in, and the older it is the better.

4 Ervum lens. *The Lentil. Lin. Sp. pl.* 1039.

Lens vulgaris. *Bauh. Pin.* 346.

The *Lentil* is a common weed in the corn-fields in France. It is an annual, and fends up feveral weak ftalks, about half a yard high, putting forth winged leaves at the joints, each being compofed of many pair of narrow lobes, and the midrib ending with a tendril. The flowers come out from the fides of the branches, two or three together on a fhort common peduncle; they are ſmall,

fmall, of the pea-kind, of a pale purple colour, contain ten ftamina each, nine of which
are united, and the tenth ftands off. The
germen is oblong, and becomes a jointed,
taper pod, containing three or four round,
convex feeds.

Lentils are a ftrong, flatulent food, very
hard of digeftion, and therefore are feldom
ufed now but to boil in foups, in order to
thicken them.

5 LOTUS edulis. *Incurved-podded Birds-
foot Trefoil. Lin. Sp. pl.* 1090.

Lotus pentaphyllos, filiqua cornuta.
Bauh. Pin. 332.

It fends forth feveral trailing ftalks about
a foot long, furnifhed at their joints with
trifoliate, roundifh, fmooth leaves, having
oval ftipulæ. The flowers come fingly from
the fides of the ftalks, on long peduncles,
with three oval floral leaves, the length of
the flower; the latter is fmall, yellow, and
is fucceeded by a thick arched pod, having
a deep furrow on its outfide.

The plant is an annual, and a native of
feveral parts of Italy, where the inhabitants
eat the young pods as we do Kidney Beans.

6 LOTUS tetragonolobus. *Square-podded
Pea. Lin. Sp. pl.* 1089.

Lotus ruber, filiqua angulofa. *Bauh.
Pin.* 332.

3

This

This is a native of Sicily, and being rather an ornamental plant, has been long cultivated in the Englifh gardens. It is an annual, fending out feveral decumbent ftalks, about a foot long, furnifhed with dark green, trifoliate leaves, having two appendages at the bafe of their footftalks. The flowers fpring alternately from the joints of the ftalks, and each is fupported on a long peduncle; they are of the pea kind, of a dark red colour, and are fucceeded by long taper pods, having four longitudinal, leafy membranes, which render them fquare.

The green pods of this plant were formerly gathered, and drefſed in the manner of Kidney Beans, and are ufed fo ftill in fome of the northern counties of England; but they are coarfe, and not very agreeable to fuch as have been accuftomed to feed upon better fare.

7 LUPINUS albus. *White Lupine. Lin. Sp. pl.* 1015.

Lupinus fativus, flore albo. *Bauh. Pin.* 347.

This grows naturally in the Levant, is an annual, and puts forth a thick, erect ftalk, near two feet high, which branches towards the top, and is furnifhed with compounded leaves, made up of feven or eight oblong, greyifh-green, hairy lobes, joined to the top of the footftalk by their tails, and are

covered

covered with a filvery down. The branches
are terminated by loofe fpikes of white
flowers, having little or no peduncles; they
are of the pea kind, and are followed by
ftraight, compreffed, hairy pods, about three
inches long, each containing five or fix flattifh white feeds, having a fcar like a navel.

This plant is cultivated in fome parts of
Italy, as an efculent pulfe, but the feeds
have a bitter difagreeable flavour.

8 PHASEOLUS vulgaris. *Kidney Bean.*
Lin. Sp. pl. 1016.
Smilax hortenfis five Phafeolus major.
Bauh. Pin. 339.
The *Common Kidney Bean* is a native of
both the Indies, and is well known by being
cultivated almoft all over Europe. The varieties of it are very numerous, but to defcribe them all would anfwer no good purpofe, as many of them are very ordinary,
and not fit for the table. Thofe generally
intended for an early crop are the *White
Dwarf,* the *Black Dwarf,* and the *Liver-
coloured;* but the moft valuable ones, though
but feldom cultivated, are the *Scarlet-blof-
fomed,* with purple feeds fpotted with black,
and the *White-bloffomed,* with white feeds.

9 PISUM fativum. *The Pea. Lin. Sp.
pl.* 1026.
This is a native of England, and, like all
plants

plants that are in conftant cultivation, is now run into many varieties. The names of thofe generally raifed for the table are,

1 The *Golden Hotfpur.*
2 The *Charlton.*
3 The *Reading Hotfpur.*
4 *Mafter's Hotfpur.*
5 The *Effex Hotfpur.*
6 The *Dwarf Pea.*
7 The *Sugar Pea.*
8 The *Spanifh Morotto.*
9 The *Nonpareil.*
10 The *Dwarf Sugar.*
11 The *Sickle Pea.*
12 The *Marrowfat.*
13 The *Rofe,* or *Crown Pea.*
14 The *Rouncival.*

10 PISUM Americanum. *Lord Anfon's Pea.*

The feeds of this Pea were brought to England by Lord Anfon's cook, who col-lected them when they were at Cape Horn, in South America. It hath weak trailing ftalks, furnifhed with compound leaves, that have two lobes on each footftalk; thofe be-low are fpear-fhaped, and fharply indented on their edges, but the upper ones are fmall and arrow-pointed. The flowers are blue, and come out by three or four on a common peduncle, and are fucceeded by taper pods, containing feveral fmall peas, about the fize of Tares.

Thefe Peas are not valuable for their flavour, being inferior to any of our cul-tivated forts, but they proved very beneficial to the failors in their voyage, who when they met with them were greatly afflicted

X   with

with the fcurvy, and ftood much in need of
·fome forts of vegetables.

11 Pisum maritimum.   *Sea Pea.*   *Lin.*
Sp. pl. 1027.

Pifum marinum.   *Raii Hift.* 892.

The *Sea Pea* grows wild on our fea-coaft,
where its roots penetrate to a confiderable
depth, and alfo fpread in various directions
for feveral feet juft under the furface. The
ftalk is angular, ufually lodges on the
ground, and grows to near a yard in length.
The leaves on the main ftalks ftand by pairs,
but thofe on the branches are pinnated,
having three or four pair of oval lobes
each, and their midrib is terminated with a
branched tendril. The flowers finifh the
ftalks in clufters of eight or ten on a common peduncle; they are fmaller than thofe
of the garden Pea, and are of a pale purple,
tinged in the middle with a bluifh purple.
The Peas have a bitterifh, difagreeable tafte,
and therefore whilft more pleafant food is
to be obtained, thefe are rejected; but in
times of fcarcity they have been the means
of preferving thoufands of families from
perifhing, the delicacy of flavour at fuch
times weighing little with a keen appetite.
Both *Stowe* and *Camden* relate, that in the
year 1555, being a year of great dearth, the
people collected large quantities of thefe
peas between Orford and Aldborough, in
  5   Suffolk,

Suffolk, upon a barren heath, where even grafs would not grow; and as they never had obferved any fuch plant as this there in the time of their fullnefs, when the eye is carelefs, they attributed their fpringing up then as a pure miracle, to keep the poor from ftarving, though in all probability they had been growing thereabouts for centuries before.

12 VICIA faba. *The Broad Bean.* Lin. *Sp. pl.* 1039.

The *Common Broad Bean* is a native of Egypt, and like the Pea is now run into many varieties, which have their diftinguifhing appellations among the gardeners, as,

1 The *Mazagan*.	5 The *Sandwich*.
2 The *Portugal*.	6 The *Toker*.
3 The *Small Spanifh*.	7 The *Windfor*, and
4 The *Broad Spanifh*.	8 The *Muntford*.

Which laft is a fmall fort of the *Windfor*. The only variety taken notice of by Linnæus is the *Horfe-bean*, and even this now is run into many variations. Thefe are not eaten in England, but our Merchants fhip them for Africa, where they are bought as fupport for the flaves in their voyage to the Weft Indies.

The diftilled water of the flowers of *Beans* has been held in great efteem as a good cofmetic among the Ladies.

S E C T. II.

Pods and Seeds of Trees.

1 CASSIA fiftula. *Sweet Caffia*, or
  Pudding-pipe Tree.
2 Ceratonia Siliqua. *Carob*, or *St. John's
  Bread.*
3 Coffea Arabica. *Arabian Coffee.*
4 Coffea occidentalis. *American Coffee.*
5 Cytifus cajan. *Pigeon Pea.*
6 Epidendrum vanilla. *Sweet-fcented Vanilla.*
7 Hymenæa courbaril. *Baftard Locuft
  Tree.*
8 Tamarindus indica. *The Tamarind.*

1 CASSIA fiftula. *Sweet Caffia. Lin.
Sp. pl.* 540.
  Caffia fiftula Alexandrina. *Bauh. Pin.*
405.
. This is a native of Alexandria, and both
the Indies. It is a large tree, fometimes
reaching to fifty feet high, having a thick
trunk, which divides into many branches,
furnifhed with winged leaves, compofed of
five pair of fmooth, fpear-fhaped lobes.
The flowers come forth in long fpikes at
the ends of the branches, fuftained on long
peduncles;

peduncles; they are yellow, and each confifts of five large concave petals, furrounding
ten ftamina, the three lower of which are
long, and tipped with arched, beaked, gap-
ing fummits. In the centre is feated a long
taper germen, which becomes a pod divided
into many cells by tranfverfe partitions, and
is from one to two feet long, with a feam
running the whole length on one fide, and
the mark of one on the other. The partitions of the pod are covered with a black
fweet pulp, which is agreeable, but purgative.

There are two forts of *Caffia* kept in the
fhops, one brought from the Eaft Indies,
and the other from the Weft. The pods of
the latter are moftly large, thick rind, and
contain a naufeous pulp; thofe of the former are generally fmaller, fmoother, the
pulp blacker, and of a fweet and more pleafant tafte. The pulp is the part ufed in
medicine, and is frequently ordered either
alone or in compofition againft coftive habits
of body. The young tender pods, when
about the fize of fmall Kidney Beans, are
preferved with fugar in the Indies, and pod,
pulp and all, eaten in the above diforders.

2 CERATONIA filiqua. *Carob-tree. Lin.
Sp. pl.* 1513.
Siliqua edulis. *Bauh. Pin.* 400.
This tree grows naturally in many places

of the Levant, and alfo in fome parts of Spain and Italy, as is afferted, but this feems doubtful. It is male and female in diftinct trees, and grows to a large fize. The body is covered with an afh-coloured bark, and the branches are furnifhed with winged, oval-lobed leaves, terminated by an odd one. The male flowers have no petals, but each confifts of a large calyx, cut into five parts, and contains five long, awl-fhaped ftamina, tipped with large twin fummits. The female flowers alfo have no petals, but a flefhy germen fituated within the receptacle, which becomes a long, flefhy, compreffed pod, divided into feveral cells, each containing one large, roundifh, compreffed feed.

Thefe pods are thick, mealy, and of a fweetifh tafte, and are eaten by the poor inhabitants in times of fcarcity; but they are apt to pain the bowels, and prove purgative. They are called *St. John's Bread,* from an affertion of fome writers on Scripture, that thefe pods were the *Locufts* St. John eat with his honey in the Wildernefs. But Dr. Haffelquift has fufficiently refuted this wild conceit, he obferving that the animals, called *Locufts,* are plentifully eaten to this day in the places where St. John was, and it is not to be doubted but they were the food he is faid to have been fupported with.

3 COFFEA Arabica. *Arabian Coffee.*
Lin. Sp. pl. 245.

This is fuppofed to be a native of Arabia
Felix, where it is greatly cultivated; It is
but a fmall tree, feldom growing above
fifteen or eighteen feet in its natural ftate,
but the planters crop it, and fcarcely fuffer
it to reach fix. The ftem is covered with a
light brown bark, and the branches diverge
oppofite each other in an horizontal di-
rection; they are furnifhed with numerous
beautiful, fharp-pointed leaves, fomewhat
refembling thofe of the Sweet Chefnut. The
flowers are produced in clufters at the bafe
of the leaves, fitting clofe to the branches,
and each confifts of a funnel-fhaped petal,
having a cylindrical tube, and is cut at the
brim into five parts. They are white, have
a moft grateful fmell, but are of fhort duration. In the tube of the flower are inferted five awl-fhaped ftamina, and below
is a roundifh germen, which turns to an
oval berry, containing two oval feeds, which
are plain on one fide, and convex on the
other.

4 COFFEA occidentalis. *American Coffee.*
Lin. Sp. pl. 246.

Pavetta foliis oblongo-ovatis oppofitis,
ftipulis fetaceis. *Browne's Jam.* 142. *t.* 6.
f. 1.

This is a native of America, and it differs

X 4   from

from the former in the flower being cut into four parts, and in the berry containing but one feed.

Of thefe two forts of *Coffee*, the Arabian is to be preferred, as having the moft grateful flavour when infufed. They are both of a drying nature, and are therefore good in diforders of the head, proceeding from fumes and moifture. They alfo promote digeftion, and remove drowfinefs, but their frequent ufe is forbidden in thin hectic conftitutions, as they are apt to dry the nerves of fuch perfons, and bring on tremblings.

5 CYTISUS cajan. *Pigeon Pea. Lin. Sp. pl.* 1041.

Laburnum humilius, filiqua inter grana et grana juncta, femine efculento. *Sloane's Jam.* 139. *Hift.* 2. *p.* 31.

This is a native of India, but is now cultivated in almoft all the American iflands. It is a fhrubby tree, and feldom exceeds ten feet in height. The leaves ftand three together upon a common footftalk, two of which are feffile and oppofite, and the middle one is protruded beyond them. They are woolly, and nearly lance-fhaped. The flowers come out in racemi from the fides of the branches, are of the pea kind, of a deep yellow colour, about the fize of the common Laburnum, and are fucceeded by

hairy,

hairy, fickle-fhaped pods, about three inches long, ending in an acute point. Thefe are of a ruffet colour, and each contains feveral roundifh kidney-fhaped feeds, which have a flight aftringent tafte; but when boiled they afford an agreeable and nutritious food.

This tree is of great utility to the inhabitants of the Weft Indies, for it not only furnifhes them with a wholefome diet, but alfo affords a conftant fupport for their Pigeons, whence the name of *Pigeon Pea*.

6 EPIDENDRUM vanilla. *Sweet-fcented Vanilla. Lin. Sp. pl.* 1347.

Epidendrum fcandens, foliis eliptico ovatis nitidiffimis fubfeffilibus, inferioribus claviculis jugatis, fuperioribus oppofitis. *Browne's Jam.* 326.

This is a parafitical plant, and grows naturally in both the Indies, where it climbs up the bodies of trees by means of its fpiral tendrils, fhooting its fibres into the bark in manner of our ivy. The leaves are oblong-heart-fhaped, of a bright green colour on the upper fide, of a paler one on the other, and have feveral prominent veins running through them. They are produced alternately at every joint, and have no foot-ftalks. The flowers are of a yellowifh-green colour, mixed with white; they have no calyx, but each is compofed of five fpreading, oblong petals, included in a fheath,

fitting

fitting upon the germen. Thefe have top-
fhaped nectariums on their backs, and their
brims are oblique, and bifid, except the upper one, which is fhort and trifid. The
germen is flender, twifted, and feated under
the flower, fupports a fhort ftyle, having two
ftamina fitting upon it, is crowned by an
obfolete ftigma, and is faftened to the upper
lip of the flower. It fwells to a long, taper,
flefhy pod, including many fmall feeds.

Thefe pods are fix or feven inches long,
of a reddifh colour, wrinkled, and very oily.
They contain a pulp that fmells like Balfam
of Peru, of an aromatic tafte, and is made
ufe of by the manufacturers of Chocolate to
give it a flavour. As thefe pods furnifh an
article of trade, the inhabitants collect them
juft as they turn ripe, and in order to preferve them for fale, they firft lay them in
heaps for two or three days to ferment, after
which they are fpread in the fun, and when
about half dried, they flat them, and rub
them over at the fame time with the oil of
Palma Chrifti. This done, they are again
expofed to the fun, and being once more
rubbed with the fame oil, they are covered
over with the leaves of the Canna Indica,
and are then properly prepared for market. *Vanillas* are deemed cordial, good to
ftrengthen the ftomach, help digeftion, dif-
fipate wind, and to fortify the brain.

7 HYMENÆA

7 HYMENÆA, courbaril. *Baſtard Lo-
cuſt-tree. Lin. Sp. pl.* 537.

Arbor ſiliquoſa ex qua Gummi Elemi,
Bauh. Pin. 404.

This is a large tree, growing naturally in
the Spaniſh Weſt Indies. The trunk is
covered with a light aſh-coloured bark, is
often more than ſixty feet high, and three
in diameter. The branches are furniſhed
with dark green leaves, which ſtand by pairs
on one common footſtalk, diverging from
their baſe in manner of a pair of ſhears, when
opened. The flowers come out in looſe
ſpikes at the ends of the branches, and are
yellow, ſtriped with purple. Each conſiſts
of five petals, placed in a double calyx, the
outer leaf of which is divided into five parts,
and the inner one is cut into five teeth at its
brim. In the centre are ten declining ſtamina, longer than the petals, ſurrounding
an oblong germen, which becomes a thick,
fleſhy, brown pod, four or five inches long,
and one broad, with a ſuture on both edges,
and includes three or four purpliſh ſeeds,
ſomewhat of the ſhape of Windſor Beans,
but ſmaller.

The ſeeds are covered with a light brown
ſugary ſubſtance, which the Indians ſcrape
off and eat with great avidity, and which is
very pleaſant and agreeable.

At the principal roots under ground is
found collected in large lumps a yellowiſh-
red,

red, tranfparent gum, which diffolved in rectified fpirit of wine affords a moft excellent varnifh, and is the gum *Anime* of the fhops, not the gum *Elemi* *.

8 TAMARINDUS indica. *The Tamarind.* *Lin. Sp. pl.* 48.

Siliqua Arabica, quæ Tamarindus. *Bauh. Pin.* 403.

The *Tamarind* is a pretty large tree, growing naturally in both the Indies, but thofe in the Eaft produce the beft and largeft fruit. The trunk is covered with a brown bark, and fpreads into many branches at the top, plentifully furnifhed with long, flender, pinnated leaves, the lobes of which are very narrow, and not above half an inch long; thefe are of a bright green colour, a little hairy, and fit clofe to the midrib. The flowers are produced from the fides of the branches, in fmall clufters of fix or eight together upon a common peduncle. Each has a calyx compofed of five equal, oval leaves, furrounding five reddifh petals, fo difpofed as to refemble a pea-flower, but they contain only three awl-fhaped ftamina, feated in the finufes of the calyx, and are arched towards the upper petal. The germen is

* This gum has been generally, though wrongfully, fuppofed to be the gum *Elemi*, but that is the gum of a tree called *Amyris Elemifera*, and is of a much paler colour than the *Anime*.

an

an oblong - oval, and supports a slender ascending style, crowned by a single stigma.

The pods when fully grown are from three to six inches long, and filled with a stringy, acid pulp, surrounding several hard seeds. This pulp is of a cooling laxative nature, is good to quench thirst, allay immoderate heat, and is an ingredient in the Lenitive Electuary of the shops.

CHAP.

CHAP. IX.

ESCULENT GRAIN and SEEDS.

SECT. I.

The various Sorts of Wheat.

LINNÆUS comprehends all the forts of *Wheat* at prefent cultivated, under the fix following fpecies :

1 Triticum æftivum. *Summer,* or *Spring Wheat.*
2 Triticum hybernum. *Winter,* or *Common Wheat.*
3 Triticum turgidum. *Short thick-fpiked Wheat.*
4 Triticum Polonicum. *Poland Wheat.*
5 Triticum fpelta. *German,* or *Spelt Wheat.*
6 Triticum monococcum. *St. Peter's Corn.*

Cultivation has produced fo many varieties from thefe fix fpecies, that the moft curious examiner cannot fix with certainty to which of them they individually belong ; but fuch as are not to be doubted, fhall be mentioned after the defcription of each fpecies.

1 TRITICUM

1 TRITICUM æftivum. *Spring Wheat.*
Lin. Sp. pl. 126.

Triticum radice annua, fpica glabra
ariftata. *Roy. lugdb.* 70.

This hath four flowers in a calyx, three
of which moftly bear grain. The calyces
ftand pretty diftant from each other on both
fides a flat, fmooth receptacle. The leaves
of the calyx are keel-fhaped, fmooth, and
they terminate with a fhort arifta. The
glumes of the flowers are fmooth and bellying, and the outer leaf of three of the
glumes in every calyx is terminated by a
long arifta, but the three inner ones are
beardlefs. The grain is rather longer and
thinner than the common Wheat. It is
fuppofed to be a native of fome part of Tar-
tary. The farmers call it *Spring Wheat,*
becaufe it will come to the fickle with the
Common Wheat, though it be fown in February or March. The varieties of it are:

Triticum æftivum fpica et grana rubente. Spring
Wheat, with a red fpike and grain.
Triticum æftivum rubrum, fpica alba. Red Spring
Wheat, with a white fpike.
Triticum æftivum, fpica et grana alba. Spring
Wheat, with a white fpike and grain.

2 TRITICUM hybernum. *Common Wheat.*
Lin. Sp. pl. 126.

Triticum

Triticum radice annua, ſpica mutica. *Roy. lugdb.* 70.

This hath alſo four flowers in a calyx, three of which are moſtly productive. The calyces ſtand on each ſide a ſmooth, flat receptacle, as in the former ſpecies, but they are not quite ſo far aſunder. The leaves of the calyx are bellying, and ſo ſmooth, that they appear as if poliſhed, but they have no ariſta. The glumes of the flowers too are ſmooth, and the outer ones near the top of the ſpike are often tipped with ſhort ariſta. The grain is rather plumper than the former, and is the ſort moſt generally ſown in England, whence the name of *Common Wheat*. Its varieties are:

> *Triticum hybernum, ſpica et grana rubente.* Common Wheat, with a red ſpike and grain.
>
> *Triticum hybernum rubrum, . ſpica alba.* Common Red Wheat, with a white ſpike.
>
> *Triticum hybernum, ſpica et grana alba.* Common Wheat, with a white ſpike and grain.

3 TRITICUM turgidum. *Thick-ſpiked Wheat. Lin. Sp. pl.* 126.

Triticum radice annua, glumis villoſis. *Roy. lugdb.* 70.

This ſpecies is eaſily diſtinguiſhed from

either

either of the former, for though it has four flowers in a calyx after the manner of them, yet the whole calyx and the edges of the glumes are covered with foft hairs. The calyces too ftand thicker on the receptacle, which make the fpike appear more turgid. Some of the outer glumes near the top of the fpike are terminated by fhort ariftæ, like thofe of the Common Wheat. The grain is fhorter, plumper, and more convex on the back, than either of the former fpecies. Its varieties are numerous, and have various appellations in different counties, owing to the great affinity of feveral of them. Thofe moft eafily to be diftinguifhed are:

Triticum turgidum conicum album. White Cone Wheat.
Triticum turgidum conicum rubrum. Red Cone Wheat.
Triticum turgidum ariftiferum. Bearded Cone Wheat.
Triticum turgidum, fpica multiplici. Cone Wheat, with many ears.

The third variety is what the farmers call Clog Wheat, Square Wheat, and Rivets. The grain of this is remarkably convex on one fide, and when ripe the awns generally break in pieces and fall off. This fort is very productive, but it yields an inferior flour to what the former two fpecies do.

4 Triticum Polonicum. *Poland Wheat.* *Lin. Sp. pl.* 127.

This has some resemblance to the turgidum, but both grain and spike are longer. The calyx contains only two flowers, and the glumes are furnished with very long aristæ. The teeth of the midrib are bearded. As this sort is seldom sown in England, there is no telling what varieties it produces.

5 Triticum spelta. *Spelt Wheat.* *Lin. Sp. pl.* 127.

Zea dicoccos vel spelta major. *Bauh. Pin.* 22.

At first view this has a great resemblance to Barley, but it has no involucrum. The calyx is truncated, that is, it appears as if the ends were snipped off, and it contains four flowers, two of which are hermaphrodite, and the glumes bearded, but the intermediate ones are neuter. There are two rows of grain as in Barley, but they are shaped like Wheat. It is much cultivated in France, Germany, and Italy, but neither the native place of this, nor of the former three species is yet known.

6 Triticum monococcum. *St. Peter's Corn. Lin. Sp. pl.* 127.

Zea Briza dicta sive monococcos germanica. *Bauh. Pin.* 21.

This

This has three flowers in each calyx, alternately bearded, and the middle one neuter. The spike is shining, and has two rows of grain in the manner of Barley. Where it grows naturally is not known, but it is cultivated in Germany, and in conjunction with Spelt Wheat is there made into bread, which is coarse, and not so nourishing as that made of Common Wheat. Malt made of any of our Wheats is often put into Beer, and a small quantity of it will give a large Brewing a fine brown, transparent tincture.

Before I quit this article of Wheat, I shall make an observation or two that may prove of some benefit to the generality of Farmers. The common allowance of seed to sow an acre, is not less than three bushels, a quantity, as Miller observes, which is certainly too much, but not perhaps altogether for the reasons he gives. If the husbandman has ten coombs per acre, for his three bushels of seed, he thinks he has had an excellent crop, nor does he set himself about reflecting how much missed coming to perfection. Now if all the grain he sowed, vegetated, and produced only two tolerable good ears each, and each ear contained only forty grains, (which is rating them full low) the produce of one grain sown would be 80, and the increase from the three bushels would be 240 bushels, or

Y 2

60 coombs;

60 coombs ; confequently, when he reaps but 10 coombs, he has the profit of only half a bufhel of his feed. It ftands the farmer in hand then to be careful about fowing his feed-corn, and not throw it away to birds and other vermin, and which he frequently does by fowing it too late. In order to prevent the ravages of thefe creatures, he ought to have all his Wheat into ground by the end of October at longeft, before the birds find a fcarcity of food ; for while there remains any part of the laft year's offal on the fields, they will not trouble themfelves much about the new fown grain; but as foon as they feel themfelves pinched, they repair by flights to the frefh fown lands, and pick up all they can poffibly get at ; and though the feeds in general may have vegetated, yet if they be not ftrongly rooted, they make little difficulty of pulling them up by their leaves, and then twitch off the grain. Several forts of birds are dexterous at this bufinefs; but Larks in particular are quite adepts at it ; a fmall parcel of them will foon make a place as bare as it was before fown. Now this wafte never happens when there is plenty of food for thefe animals; nor can it be performed when the corn is much advanced, it then requiring more than their ftrength to draw it up, fo that if it be fown in time, and before thefe

creatures

creatures are diftreffed, it fuffers little or nothing, but from the feverity of hard feafons. From what has been obferved it muft appear evident, that a much lefs quantity of feed fown early, properly fcattered, and well covered, will be productive of as large a crop as the ufual allowance is; and probably a larger, for the grains being lefs liable to be difturbed by the birds when ftriking root, and their roots ftanding more diftinct, they will be better fupplied with nourifhment, enabled to fupport their ftems, and bring their feed to greater perfection.

S E C T. II.

Oats, Barley, and Rye.

1 AVENA fativa. *Manured Black Oat.*

———— *alba.* Manured White Oat.

2 Avena nuda. *Naked Oat,* or *Pilcorn.*

3 Hordeum vulgare. *Common Barley.*

———— *cælefte.* Siberian Barley.

4 Hordeum diftichon. *Long-eared Barley.*

———— *nudum.* Naked Barley.

5 Hordeum hexaftichon. *Big,* or *Square Barley.*

Y 3

6 Hor-

6 Hordeum zeocriton. *Battledore,* or
    Sprat Barley.

7 Secale cereale. *Common Rye.*
  ——— *vernum.* Spring Rye.

   1 Avena fativa. *The Oat. Lin. Sp.*
pl. 118.

   Avena nigra. *Bauh. Pin.* 23.

The *Oat* was found growing wild by
Lord Anfon in the ifland of Juan Fernandez,
at the back of the coaft of Chili, in the
South Sea ; but probably it never was natural to this place, but had been dropped
by the Spaniards, who had been here before
Anfon. In Scotland, and fome of the nor-
,thern counties of England, *Oats* form the
chief bread of the inhabitants. They are
much ufed likewife in Germany; but in
Norway, *Oat-bread* is a luxury among the
common people, for they fpare the grain
by mixing Fir-bark with it, and grinding
both into meal. And they do this not only
in times of fcarcity, but alfo when *Oats* are
plentiful, that they may be inured to it
when the latter fail them. The Fir gives
the bread a bitterifh tafte; and therefore
lately they have generally fubftituted Elm-
bark for it, which they find much pleafanter.
Oats are very nutritive, and eafy of digeftion,
to fuch as feed conftantly upon them.

The White Oat is only a variety of the
Black, and though the former are generally
                       preferred

preferred for feeding horses, yet it has been found on some fair trials, that the latter are the best for this purpose, and that such horses as are kept with the Black Oat, appear most healthy, and fullest of spirits.

2 AVENA nuda. *Naked Oat.* *Lin. Sp. pl.* 118.

This is sometimes found in our cornfields, and is therefore supposed to be natural to England. It so much resembles the Tartarian Oat, in its manner of growth and general appearance, that it may easily be mistaken for it by any one not well skilled in plants. The difference is, this has three flowers in a calyx, whereas the Tartarian has only two; and the seed of the nuda lies bare in the husk in the manner of Rye; but that of the Tartarian is enwrapped in the glume. In former ages this was the chief *Oat* cultivated here, for the seeds being naked was a great inducement to its propagation, before the method of husking the Common Oat became general, as when they were boiled, they turned for the most part into flour.

3 HORDEUM vulgare. *Common Barley.* *Lin. Sp. pl.* 125.

Hordeum polystichon vernum. *Bauh. Pin.* 22.

This is the *Barley* most generally cultivated. It has three or four rows of flowers,

two

two of which are erect, and stand in a regular order. They are all hermaphrodite,
and bearded. The skin which covers the
feed is very thin, and consequently it is a
good fort for the maltster. *Barley* is less
nourishing than Wheat, apt to purge the
body, and therefore is not made into bread
here, but when the latter becomes too dear
for the pockets of the common people. In
the Greek islands, Barley-bread is much in
use; this and dried Figs being the principal
food of the Monks, the same as Wheaten
bread and cheese are here. In Scotland too
the poor people eat frequently of Barley-
bread. In many parts of India this grain is
much cultivated for their cattle, the inhabitants making the meal into dough, which
they form into balls, and give them to their
Oxen and Camels. Its native place of
growth is not known.

4 HORDEUM distichon. *Long - eared
Barley. Lin. Sp. pl. 125.*
This is the *Barley* generally cultivated in
Norfolk and Suffolk. The ears are very
long, and the grains are regularly ranged in
a row on each side the receptacle. They are
angular, and have a very thin skin, which
last circumstance renders this fort also very
proper for malting. The French and Pearl
Barley of the shops are said to be prepared
from this species, but as there is little difference

ference between the feeds óf this and the
former, I imagine they are both promifcuoufly ufed for this purpofe. The Pearl-
barley is prepared in Holland and Germany,
by firft fhelling the grain, and then grinding
it into round granules, which gives them a
pearly whitenefs. This boiled is very foft
and lubricating, and is either drank alone
to flake thirft, and to obtund acrimonious
humours, or it is ordered in emulfions. In
Scotland they prepare a deal of both forts,
and they are there boiled in broths to
thicken them,

5 HORDEUM hexaftichon. *Square Barley.*
Lin. Sp. pl. 125.

This goes by the feveral names of *Winter
Barley, Square Barley, Bear, Big,* and *Clog
Barley.* The flowers are all bearded, and
ranged in fix rows fo equally, as to form a
perfect fix-fided figure. In many parts of
Scotland they feldom cultivate any fort but
this, it being more hardy than the reft, and
the ears there come to a very large fize, but
the fkin being rather thickifh, the grain is
not fo good for malting as either of the
former. In Switzerland, and alfo in fome
of the Provinces of Germany, they make
bread of this, Spelt Wheat, and Oats, all
mixed together. In Egypt, where they
fow no Oats, they cultivate this as food for
their horfes.

3   6 HORDEUM

6 HORDEUM zeocriton. *Sprat Barley.*
Lin. Sp. pl. 125.

Zeocriton five Oriza germanica. *Baub.*
Pin. 22.

This has two regular rows of feed, one on
each fide the midrib, the fame as the *difti-*
chon, but the ear is fhorter and broader, the
awns are very long, the grains are clofer
crouded together, and when ripe they diverge fo as to caufe the awns to fpread very
wide, and give the idea of a Battledore;
whence the name of *Battledore Barley.* The
grain is angular like the common Barley,
but it is rather fhorter, and has a thicker
fkin, fo is not fo eligible for malting. It
generally yields plentifully to the grower,
but the firaw is fo coarfe, that cattle will
feldom eat it, for which reafon the farmers
are not fond of cultivating this fort. The
native country of any of thefe three laft
fpecies is not known.

7 SECALE cereale. *Common Rye. Lin.*
Sp. pl. 124.

Secale hybernum vel majus. *Baub. Pin.*
23.

This is a native of the ifland of Candia,
but has been cultivated in England for
many ages. About a century paft, *Rye* made
the principal bread of the common inhabitants here, but it was black, clammy,
very detergent, and confequently lefs nourifhing

rifhing than Wheat. It is ftill ufed in
Wales, in conjunction with the latter, and
in fome parts of Sweden and Norway, the
poor people feed on little elfe, Wheat-bread
being moftly preferved for feafts and weddings.

SECT. III.

Mifcellaneous Grain and Seeds.

1 COIX lacryma Jobi. *Job's Tears.*
2 Cynofurus coracanus. *Indian Cock's-Foot Grafs.*
3 Feftuca fluitans. *Flote Fefcue Grafs.*
4 Holcus forghum. *Guinea Corn,* or *Indian Millet.*
5 Holcus faccharatus. *Indian Reed Millet.*
6 Nymphæa nelumbo. *Egyptian Bean.*
7 Oryza fativa. *Rice.*
8 Panicum miliaceum. *Common Millet.*
9 Panicum Italicum. *Indian Millet.*
10 Phalaris canarienfis. *Canary Grafs.*
11 Polygonum fagopyrum. *Buck Wheat.*
12 Quercus efculus. *Cut-leaved Italian Oak.*
13 Quercus phellos. *Carolinian Willow-leaved Oak.*
14 Sefamum orientale. *Eaftern Fox-glove.*
15 Sefamum Indicum. *Indian Fox-glove.*
16 Sinapis nigra. *Black Muftard.*

17 Sinapis

17 Sinapis arvenfis. *Charlock.*
18 Zea mays. *Turkey,* or *Indian Wheat.*
19 Zizania aquatica. *Water Zizania.*

1 Coix lacryma Jobi. *Job's Tears.* Lin.
Sp. pl. 1378.
Litholpermum arundinaceum. *Bauh. Pin.*
258.

This is a native of both the Indies. It is a perennial, and fends up two or three crooked ftalks, about two feet high, having a long graffy leaf at every joint, at the bafe of which come out the fpikes of flowers, on fhort footftalks. The fpikes are compofed of all male flowers, and juft below them are two or three females. The male has a bivalve, hufky calyx, and a bivalve glume, containing three flender ftamina, tipped with oblong, four-cornered fummits. The female flower is alfo compofed of a bivalve calyx and glume, and contains an oval germen, which becomes a hard, fmooth, roundifh feed, nearly like that of Gromwel.

This plant is cultivated in Spain and Portugal, for the ufe of the poor inhabitants in the time of fcarcity, the feeds being then ground, and made into a coarfe fort of bread. As they are hard and of different colours, they are often perforated by the negroes, ftrung upon filk, and then worn for necklaces.

2 CYNOSURUS

2 CYNOSURUS coracanus. *Indian Cock's-foot Grafs.* *Lin. Sp. pl.* 106.

Gramen Dactylon Ægyptiacum. *Bauh. Pin.* 7.

This is an annual, and a native of India. It hath woolly graffy leaves, among which rife the ftems, not more than three or four inches high. Thefe are flattifh, erect, and terminated by four (fometimes fix) linear fpikes, that fpread in the form of a crofs. The flowers are all hermaphrodite, feveral ftanding together in a bivalve, hufky calyx, and each has a bivalve glume, containing three flender ftamina, and two hairy reflexed ftyles. The germen is fmall, and top-fhaped.

The feeds are near as large as fmall Millet, and are ufed by the inhabitants for the fame purpofes that Millet is.

3 FESTUCA fluitans. *Flote Fefcue Grafs.* *Lin. Sp. pl.* 111.

Gramen aquaticum fluitans, multiplici fpica. *Bauh. Pin.* 2.

This grows very common by ditches, and almoft all moift places in England. It hath a creeping root, which fends forth feveral curved ftalks, a little flatted towards the bafe; thefe are terminated by long panicles, which are very much branched when the plant grows in the water, or on a very moift place; but in drier fituations the panicles are

are scarcely branched at all. They are of a silvery green colour, and the spiculæ are round, linear, and beardless. The flowers are hermaphrodite, and several of them are common to a bivalve, husky calyx. Each is composed of a bivalve glume, longer than the calyx, and contains three slender stamina, tipped with oblong summits, together with two short, reflexed styles, crowned with simple stigmata. The seed is slender, oblong, and hath a longitudinal furrow.

These seeds are not regarded here as esculent grain, but in Poland they are yearly collected, and sent into Germany and Sweden; where they are sold by the name of Manna Seeds, for the use of the table of people of the first rank, and are much esteemed for their agreeable and nourishing quality. Linnæus affirms, that the bran of this grain will kill Bots in horses, if they be kept from drinking some time before it be given them; and that the grain itself will fatten Geese sooner than any yet known; all which clearly point out the utility of Botany to a farmer; for from this common plant only, if he should be able to distinguish it, he may draw a medicine for his diseased horses, and a profitable and nourishing food for his geese. The poorer sort of people too might collect the seeds for sale as they do in Poland, for if they are so pleasant and agreeable at the tables of the

German

German and Swedish gentlemen, why should
they not be so at those of the English? The
plant grows prodigiously plentiful in most
marshes, and in those near the sea; and in
the middle of a hot day, I have seen the
spikes quite covered with a brown substance, as sweet as sugar.

4 HOLCUS *forghum.* *Guinea Corn.* *Lin.*
Sp. pl. 1484.

Milium arundinaceum, subrotundo se-
mine, Sorgho nominatum. *Bauh: Pin.* 26.

This is an annual plant, and a native of
India. It sends up thick, strong stalks, like
those of Turkey Wheat, to seven or eight
feet high, and set at their joints with large
grassy leaves, often more than two feet long,
and three inches broad in the middle, embracing the stalks with their base. The
midrib of these is very depressed on the upper surface, and prominent on the back.
The stalks are terminated with large, close,
oval panicles of chaffy flowers, some of
which are male, and others hermaphrodite
on the same panicle. The male flowers
have no glumes, but each consists of an
hairy, husky, bivalve calyx, containing three
hairy stamina, tipped with oblong summits.
The hermaphrodite flowers have a like, but
larger calyx, together with a bivalve glume,
containing three hairy stamina, and two
small styles, crowned with pencil-shaped
2   stigmata.

ftigmata. The germen is roundifh, and
becomes an oval feed, wrapped in the
glume, having a fmall arifta, the bottom of
which is brown, and the top white.

5 HOLCUS faccharatus. *Indian Reed
Millet. Lin. Sp. pl.* 1484.

Frumentum indicum, quod Milium indicum vocant. *Bauh. Theatr.* 488.

This too is a native of India, grows to the
fize of the former, and makes the like general appearance; but the panicle of this
fpreads open, the branches ftanding nearly
horizontally upon the receptacle. The calyces of the flowers too are fmooth, but the
feeds are much of the fame fize as the former; thefe vary in both with refpect to
colour, they being white, yellow, or reddifh. The ftalks of this fpecies are almoft
as copioufly ftored with a faccharine juice as
the Sugar Cane.

Both thefe plants are cultivated in Africa
by the name of Guinea Corn, and they have
been confounded as only one fort by moft
travellers. The grain is there made into
bread, and otherwife ufed, and is deemed
wholefome food. From Africa the Negroes
carried them to the Weft Indies, where they
are both fown for their ufe, and each flave
is generally allowed from a pint to a quart
per day.

6 NYMPHÆA

6 NYMPHÆA nelumbo. *Egyptian Bean.*
Lin. Sp. pl. 730.

Nymphæa foliis orbiculatis peltatis fubtus
radiatis. *Browne's Jam.* 343. Faba Ægyptia.

This is a perennial, growing naturally in
ftagnated waters, in both the Indies. It
fends forth large, orbicular leaves, which
float upon the furface of the water, and are
about half a yard diameter, having their
footftalks, which are long and prickly, inferted into their centre. From the middle
of each leaf iffue a great number of large
rays or ribs, all diverging towards the margin, breaking into many ramifications, and
making a beautiful appearance. Among the
leaves come the flowers, fupported on long
peduncles; they are large, and confift of
many deep flefh-coloured petals, difpofed in
rows, as they are in the White Water Lily.
In the middle are numerous incurved ftamina, furrounding an oval germen, which
becomes a top-fhaped feed-veffel, having
many cells, that form as many holes upon
its furface, in manner of a fand-difh, each
containing a fingle feed.

When thefe feeds are young and green,
they are boiled and eaten by the inhabitants
of India, they being then agreeable; but
when full ripe, they are hard and bitterifh.
I knew a perfon who eat many of them raw,
as they were fent from the Weft Indies, and
they made him very ill for fome time after.

Z The

The flowers of this plant are sacred in some heathen countries, and with them they adorn the altars of their temples. Often too their gods are painted sitting upon them.

The ancient writers on Botany mostly confounded this plant with the *Arum colocasia*, which caused much confusion in their accounts of both plants, and was the means of inducing many to believe that the *Faba Ægyptia* existed only in the brains of such as wrote about it. This uncertainty seems to have arisen from some affinity in the leaves of the two plants, they both being peltated, and though not exactly of a shape, yet in more remote times, when this science was very imperfect, such differences were not strictly attended to, and therefore it is probable, that those who did not see the plants in flower, mistook the one for the other; which they might easily do, as they both grow in the same kind of soil and situations.

7 Oryza sativa. *Rice. Lin. Sp. pl.* 475. *Rice* is a native of India, and is cultivated in almost every part of Asia. It is an annual, and rises to about a yard high, with broader and thicker leaves at the joints of the stalks, than those of Wheat. Each stalk is terminated by a spreading panicle, plentifully furnished with small flowers, standing singly in a bivalve chaffy calyx,

and

and having a bivalve, boat-fhaped glume, ending in a fpiral beard. The ftamina are fix, of the length of the glume, and are terminated by fummits, which fplit at their bafe. There are two hairy, reflexed ftyles, crowned with feathered ftigmata, and placed on a top-fhaped germen, which becomes an oblong comprefled feed.

This grain is the principal food of the inhabitants in all parts of the Eaft, where it is boiled and eaten either alone or with their meat. Large quantities of it are annually fent into Europe, and it meets with a general efteem for family purpofes. The people of Java have a method of making puddings of *Rice*, which feems to be unknown here, but is not difficult to put in practice, if it fhould merit attention. They take a conical earthen pot, which is open at the large end, and perforated all over; this they fill about half full with *Rice*, and putting it into a larger earthen pot of the fame fhape, filled with boiling water, the *Rice* in the firft pot foon fwells and ftops the perforations, fo as to keep out the water; by this method the *Rice* is brought to a firm confiftence, and forms a pudding, which is generally eaten with butter, oil, fugar, vinegar, and fpices. The Indians eat ftewed *Rice* with good fuccefs againft the bloody-flux, and in moft inflammatory diforders they cure themfelves with only a decoction of it. The fpirituous

Z 2

liquor,

liquor, called *Arrack*, is made from this grain.

Rice grows naturally in moist places, and will not come to perfection when cultivated, unless the ground be sometimes overflowed, or plentifully watered. The grain is of a grey colour when first reaped, but the growers have a method of whitening it, before it is sent to market. The manner of performing this and beating it out in Egypt, is thus related by Haffelquift: They have hollow iron, cylindrical pestles, about an inch diameter, lifted by a wheel worked with oxen. A person fits between the pestles, and as they rife, pushes forward the Rice, whilst another winnows, and supplies fresh parcels. Thus they continue working, until it is entirely free from chaff. Having in this manner cleaned it, they add one-thirtieth part of salt, and rub them both together, by which the grain acquires a whitenefs; then it is paffed through a fieve, to feparate the falt again from it.

In the Island of Ceylon they have a much more expeditious method of getting out the Rice, for in the field where it is reaped, they dig a round hole with a level bottom, about a foot deep, and eight yards diameter, and fill it with bundles of the corn. Having laid it properly, the women drive about half a dozen oxen continually round the pit, and thus they will tread out forty or fifty bufhels a-day.

a-day. This is a very ancient method of treading out corn, and is ftill practifed in Africa upon other forts of grain.

8 Panicum miliaceum. *Common Millet.* *Lin. Sp. pl.* 86.
Milium femine luteo & albo. *Bauh. Pin.* 26.

This is a native of India. It fends up a channelled, reed-like ftalk, to the height of about four feet, compofed of four or five joints, and furnifhed with a large grafly leaf at each, the bafe of which is covered with foft hairs, and embraces the ftalk up to the next joint. The ftalk is terminated by a large loofe panicle of green flowers, each confifting of a trivalve calyx, one part of which is very fmall, and a bivalve glume, containing three hairy ftamina, and two hairy ftyles, crowned with pencil-fhaped ftigmata. The germen is roundifh, and becomes a feed of the fame form, covered with the glume.

This plant is cultivated in moft eaftern countries, and alfo in feveral of the warm parts of Europe. The feeds vary in their colour, and are white, yellow, or blackifh. They are pretty well known here, being frequently made ufe of for puddings.

9 Panicum Italicum. *Indian Millet.* *Lin. Sp. pl.* 83.

Z 3

Panicum

Panicum Italicum five paniculâ majore. *Bauh. Pin.* 27.

This is a native of both the Indies, and grows to much the fame height as the former; but it has a compound fpike, not a panicle, and the fmaller fpikes grow in clufters, mixed with briftles, upon hairy peduncles, and a hairy midrib. The bafes of the leaves are covered with hairs. It is much cultivated in Italy, and fome parts of Germany, where they make puddings of the feeds, and alfo boil them in moft of their foups and fauces.

10 PHALARIS canarienfis. *Canary Grafs.* *Lin. Sp. pl.* 79.

Phalaris major, femine albo. *Bauh. Pin.* 28.

This is a grafs-leaved plant, and grows naturally in the Canary Iflands. It rifes to about two feet high, having crooked, channelled ftalks, with a leaf at each joint, the fheath of which embraces the ftalk to the next joint. The ftalk is terminated with an egg-fhaped, compound fpike, thickly fet with flowers, each having a bivalve, keel-fhaped calyx, of a yellowifh colour, ftriped with green, and a bivalve glume, containing three ftamina and two ftyles.

The feed is well known, being the ufual food of Canary-birds. In its native country

the

the inhabitants grind it into meal, and make a coarse sort of bread with it.

11 POLYGONUM fagopyrum. *Buck Wheat. Lin. Sp. pl.* 522.

Eryfimum cereale, folio hederaceo. *Bauh. Pin.* 27.

The *Buck Wheat* is fo often found wild in our tilled lands, that it is fuppofed to be natural here, but it is probable it was at firft introduced from Afia. It is frequently cultivated by the farmers, which makes it generally known, and therefore it will be needlefs to defcribe it. In feveral parts of Europe this conftitutes the principal food of the poor inhabitants; and in Ruffia in particular, it was formerly not only eaten by the lower clafs, but even the nobility were contented with it. Boiled and then buttered it was fuch a favourite difh of the Czar Peter the Great, that it is faid he feldom fupped on any thing elfe. This method of eating *Buck Wheat* is ftill in great efteem both in Germany and Switzerland. They make cakes and puddings of it too, and boil it in their broths and foups.

12 QUERCUS efculus. *Italian Oak. Lin. Sp. pl.* 1414.

Quercus parva five Fagus Græcorum et Efculus. *Bauh. Pin.* 420.

This fort of *Oak* grows naturally in the

fouth

fouth of France and Italy. It hath fmooth
finuated leaves, fo deeply cut, that they
appear like lobes. Their footftalks are fhort,
and fome of the finufes end in an acute
point, others in an obtufe one. The young
branches are covered with a purplifh bark,
and the acorns fit clofe to them. The latter
are long, flender, with very rough cups.

In times of fcarcity the poor people in
France collect thefe acorns, and grind them
into meal, of which they make bread. They
have a fweetifh tafte, but afford little nourifhment.

13 QUERCUS phellos. *Willow - leaved
Oak.* Lin. Sp. pl. 1412.

This is an ever-green, and a native of
Virginia. It is a very large tree, often
rifing upwards of forty feet high. The
wood is hard, tough, and coarfe. The
branches are covered with a greyifh bark,
and are garnifhed with oblong, fpear-fhaped
leaves, fomewhat like thofe of Sallow, but
of a thicker confiftence. The acorns are
oblong, and fit in very fhort cups; they are
fweeter than a Chefnut, and are much fought
after by the Indians, in order to lay up to
regale with in Winter. They likewife draw
an oil from them, which they ufe inftead of
butter, and it is little inferior to the oil of
Almonds. In America the tree goes by the
name of *Live Oak.*

14 SESAMUM

14 SESAMUM orientale. *Eastern Fox-glove. Lin. Sp. pl.* 883.

Sefamum veterum. *Bauh. Pin.* 27.

This is an annual, and grows naturally in the ifland of Ceylon, and on the coaft of Malabar. It fends up a round, hairy ftalk, about two feet high, divided into a few branches, furnifhed with oblong-oval leaves, ftanding oppofite on footftalks; they are entire on their margins, veined, and thinly covered with a few foft hairs. The flowers come out fingly at the bofoms of the leaves, upon fhort peduncles; they are white, and each has a permanent calyx, cut at the brim into five equal parts, which fpread open, and contain a petal fhaped like that of the Foxglove. In the centre of the tube are four ftamina, two fhorter than the other, and all fhorter than the petal; thefe furround an oval hairy germen, fupporting a ftyle longer than the ftamina, and crowned by a fpear-fhaped ftigma, divided into two parts. When the flower falls, the germen becomes an oblong capfule, having four cells, containing many fmall oval, compreffed feeds.

This plant is not only cultivated in Afia, but alfo in Africa, and from the latter the negroes have carried it to South Carolina, where they raife large quantities of it, being very fond of the feeds, and make foups and puddings of them, as with Rice and Millet. They parch them too over the
fire,

fire, and with other ingredients, ftew them
into a hearty food.. The feed in Carolina
is called *Oily Grain,* it yielding oil very copioufly. This when firft drawn has a warm
pungent tafte, and is otherwife not palatable,
but after being kept a year or two, the difa-
greeablenefs goes off, and it becomes mild
and pleafant, is then ufed in their fallads,
and for all the purpofes of *Olive Oil.*

15 SESAMUM Indicum. *Indian Foxglove.*
Lin. Sp. pl: 884.

This too is an annual, and a native of
fome parts of India, The ftalk rifes higher
than in the former fpecies, and the lower
leaves are cut into three divifions. The
flower refembles the other, and the grain is
eaten in India in the fame manner.

16 SINAPIS nigra. *Black Muftard. Lin,*
Sp. pl. 933.

Sinapi rapi folio. *Bauh. Pin.* 99.

This is an annual, and grows wild in
hedges, and on the borders of our fields. It
fends up a branched ftalk, three or four feet
high, furnifhed with varioufly jagged leaves
at the divifions of the branches; thofe at
the lower part refemble Turnep leaves, tho'
fmaller, but towards the top they are lefs
jagged, and nearly oval. The flowers terminate the branches in loofe fpikes; they
are yellow, and each is compofed of a calyx

5   of

of four narrow leaves, which spread open
in form of a cross, and fall off when the
flower fades; and of four roundish petals,
standing in the same manner, having four
oval glands, one on each side the stamina
and style.. In the centre are six awl-shaped
stamina, two shorter than the rest, surrounding a taper germen, which becomes a
smooth four-square pod, about an inch long,
ending in a sharp point.

This plant is cultivated for the seed, of
which that excellent and wholesome sauce,
called Mustard, is made.

17 SINAPIS arvensis. *Charlock. Lin.
Sp. pl.* 933.

Rapistrum flore luteo. *Bauh. Pin.* 95.

This is the *Common Charlock,* and it is
generally known by being a troublesome
weed among corn. It is said the *Durham*
flour of Mustard is made from the feeds of
this; but the truth of it I know not. There
is another plant called *Charlock,* or *Wallock,*
by the farmers, and grows larger than the
former. This is the *Raphanus raphanistrum,*
the calyx of which is shut, or stands upright, the flower is whitish, and the pod is
long, round, smooth, and has but one cell.
This is a more pernicious weed among corn
than the first *Charlock.*

18 ZEA

18 ZEA mays. *Indian Wheat. Lin.
Sp. pl.* 1378.

Frumentum Indicum Mays dictum.
Bauh. Pin. 25.

The *Turkey Wheat* is a native of America,
where it is much cultivated, as it is also in
some parts of Europe, especially in Italy
and Germany. There are many varieties,
which differ in the colour of the Grain, and
are frequently raised in our gardens by way
of curiosity, whereby the plant is well
known. It is the chief bread corn in some
of the southern parts of America, but since
the introduction of Rice into Carolina, it is
but little used in the northern colonies. It
makes a main part too of the food of the
poor people in Italy and Germany. This
is the sort of *Wheat* mentioned in the Book
of Ruth, where it is said that *Boaz* treated
Ruth with parched ears of corn dipped in
vinegar. This method of eating the roasted
ears of *Turkey Wheat* is still practised in the
East, they gathering the ears when about
half ripe, and having scorched them to their
minds, eat them with as much satisfaction
as we do the best flour-bread. In several
parts of South America they parch the ripe
corn, never making it into bread, but grinding it between two stones, mix it with water
in a Calabash, and so eat it.

The Indians make a sort of drink from
this grain, which they call *Cici*. This liquor

quor is very windy and intoxicating, and has nearly the taste of four Small Beer; but they do not use it in common, being too lazy to make it often, and therefore it is chiefly kept for the celebration of feasts and weddings, at which times they mostly get intolerably drunk with it. The manner of making this precious beverage, is to steep a parcel of the corn in a vessel of water, till it grows sour; then the old women, being provided with Calabashes for the purpose, chew some grains of the corn in their mouths, and spitting it into the Calabashes, empty them spittle and all into the sour liquor, having previously drawn off the latter into another vessel. The chewed grain soon raises a fermentation, and when this ceases, the liquor is let off from the dregs, and set by till wanted. In some of the islands in the South Sea, where each individual is his own lawgiver, it is no uncommon thing for a near relation to excuse a murderer, for a good drunken-bout of *Cici*.

19 ZIZANIA aquatica. *Water Zizania.* *Lin. Sp. pl.* 1408.

Arundo alta gracilis, foliis e viridi cæruleis, locustis minoribus. *Sloane's Jam.* 33. *Hist.* I. *p.* 110.

This is a reed-like plant, growing in the swampy parts of Jamaica and Virginia. The leaves are of a green-purplish colour, and the

the ftalks terminate in fpreading panicles of
male and female flowers in diftinct cups.
The male hath no calyx, but confifts of a
bivalve, equal glume, containing fix fmall
ftamina, tipped with oblong fummits. The
female alfo hath no calyx, but is compofed
of a bivalve glume, wrapped round the germen, and having a long arifta. The germen
fupports two fmall ftyles, and becomes a
fmall oblong feed.

The Indians are exceedingly fond of this
grain, and count it more delicious than
Rice. If this valuable plant were brought
into England, as is juftly obferved by a late
writer, it is probable it would fucceed well
upon fome of our low meadows, and amply
reward the pains of fuch as might cultivate it.

CHAP. X.

ESCULENT NUTS*.

1 AMYGDALUS communis. *Sweet and Bitter Almond.*
2 Anacardium occidentale. *Cashew Nut.*
3 Avicennia tomentosa. *Eastern Anacardium*, or *Malacca Bean.*
4 Corylus avellana. *Hazel Nut.*
——— *racemosa.* Cluster Nut.
——— *maxima.* Large Cob Nut.
——— *rubens.* Red Filbert.
——— *alba.* White Filbert.
5 Cocos nucifera. *Cocoa Nut.*
6 Fagus castanea. *Common Chesnut.*
7 Fagus pumila. *American Chesnut.*
8 Juglans regia. *Common Walnut.*
9 Juglans nigra. *Black Virginian Walnut.*
10 Jatropha curcas. *Indian Physic Nut.*
11 Jatropha multifida. *French Physic Nut.*
12 Pinus pinea. *Stone*, or *manured Pine.*
13 Pistacia vera. *Pistachia Nut.*
14 Pistacia narbonensis. *Trifoliate-leaved Turpentine-tree.*
15 Theobroma cacao. *Chocolate Nut.*
16 Trapa natans. *Jesuit's Nut.*

* A Nut is defined to be a hard, woody feed-veffel, inclofing a meat or kernel.

1 AMYGDALUS

1 AMYGDALUS communis. *The Al-mond-tree. Lin. Sp. pl.* 677.

Amygdalus fylveftris. *Bauh. Pin.* 441.

This grows wild in Africa, and rifes to a very large tree, fpreading its arms to a great width. Thefe put forth numerous flender branches, furnifhed with leaves nearly like thofe of the Peach. The flowers come out by pairs, and have little or no peduncles; they refemble the Peach flowers, but are of a lighter colour, and are fucceeded by dry, fkinny fruit, containing the nuts called *Al-monds*.

The *Almonds* are of two kinds, one fweet, the other bitter, yet both are promifcuoufly produced from kernels of the fame tree; nor does there appear any difference in the nuts to the eye. They both yield by ex-preffion a copious quantity of oil, which has neither fmell or any particular tafte. This oil is of a foft relaxing nature, and is given internally againft coughs, heat of urine, and inflammations. The kernels of the Sweet Almond are eaten in abundance, and about half a fcore of them peeled are faid to give relief in the heart-burn.

2 ANACARDIUM occidentale. *Cafhew Nut. Lin. Sp. pl.* 548.

This tree grows naturally in both the Indies, and is the only plant of the genus. It is rather low, feldom exceeding twenty

5   feet,

feet, but breaks into wide crooked branches, which are furnifhed with oval leaves, about the fize of thofe of the Pear-tree. The flowers are fmall, white, and come out at the fides of the branches; they have a pentaphyllous * calyx, compofed of oval, fharp-pointed leaves, and a bell-fhaped petal, cut at the brim into five fegments. In the centre are ten ftamina, and one inflexed, awl-fhaped ftyle, crowned by an oblique ftigma. The germen is roundifh, and becomes a large, yellow, oval, flefhy fruit, about the fize of a Lemon, fupporting at its apex, which is the thickeft end, a fmooth, afh-coloured nut, fhaped like a hare's kidney, and about an inch and a half long, and one broad.

The flefhy fruit is ftringy, and full of a rough, acid juice, which is ufed in America to acidulate punch. The fhell of the Nut is very hard, and the kernel, which is fweet and pleafant, is covered with a thin film; between this and the fhell is lodged a thick, blackifh, inflammable liquor, of fuch a cauftic nature in the frefh Nuts, that if the lips chance to touch it, blifters will immediately follow. The kernels are eaten raw, roafted, or pickled.

The cauftic liquor, juft mentioned, is efteemed an excellent cofmetic with the Weft India young Ladies, but they muft

* Having five leaves.

A a   certainly

certainly fuffer a great deal of pain in its application; and as fond as our English females are of a beautiful face, it is highly probable they would never fubmit to be flayed alive to obtain one. When any of the former think themfelves too much tanned by the heat of the fun, they take the *Cafhew* kernels, and gently fcrape off the thin fkins with which they are furrounded; with thefe they rub their faces all over, which caufe them immediately to fwell and grow black, but in a few days the fkin of the whole face flakes off in pieces, and in about three weeks a new one will be formed, which will be as fmooth and fair as that of a young child. I have been told by perfons who have ftood under thefe trees for fhelter in a ftorm, that by chance this liquor has dropped on their hands from fome decaying Nuts, and it has eaten the fkin nearly as quick as *aqua fortis.*

The yellow fruit is famous for curing the Brafilian negroes of diforders in the ftomach, to which they are very fubject; but they feldom ufe it voluntarily for this purpofe, as their humane mafters, when they find them much indifpofed, knowing what is good for their health, drive them to woods abounding with *Cafhew Nuts,* and leave them there either to perifh by famine, or cure themfelves. In a fhort time hunger forces them to eat plentifully of the fruit, there being
nothing

nothing elfe to be had, and in two or three weeks they are brought back again perfectly found, and fit for their cuftomary labour. The milky juice of this tree will ftain linen of a good black, which cannot be wafhed out again.

3 Avicennia tomentofa. *Eaftern Ana-cardium. Lin. Syft. Natu.* 426.

Bontia foliis fubtus tomentofis. *Jacq. Amer.* 25. Anacardium. *Bauh. Pin.* 511.

This tree is a native of both the Indies. The leaves are oblong, entire, woolly underneath, and ftand oppofite, on very fhort thick footftalks. The flowers are produced in long bunches, and each confifts of a permanent calyx, cut into five roundifh lobes, and containing a white bell-fhaped petal, having a fhort tube, with its brim cut into two lips, each of which is moftly divided into three equal oval parts. It hath four awl-fhaped erect ftamina, tipped with roundifh, twin fummits, and one erect ftyle, crowned with an acute, bifid ftigma. The capfule is tough, compreffed, fomewhat the fhape of a rhombus, and contains one large feed of the fame figure, having four flefhy gills.

Thefe feeds are faid to be the Malacca Beans formerly kept in the fhops, (but this is doubtful) the kernels of which were eaten as Almonds.

　The

The plant is the *Bontia germinans* of the Species Plantarum.

4 CORYLUS avellana. *The Hazel. Lin. Sp. pl.* 1417.

Corylus fylveftris. *Bauh. Pin.* 418.

The *Hazel* is fo common in our woods and hedges, that it muft be generally known. The different kinds of *Filberts*, fo commonly planted in gardens, are only varieties of this. Whether the Spanifh Nut be another variety is uncertain, but Miller thinks the latter is the *Corylus colurna.*

It will be needlefs to mention the manner of eating the Nuts here, but in China they put the meats into their Tea, and count they give it a more grateful flavour.

5 Cocos nucifera. *Cocoa Nut. Lin. Sp. pl.* 1658.

Palma indica coccifera angulofa. *Bauh. Pin.* 502.

This is a fpecies of Palm, growing naturally in the Eaft Indies, but it is much cultivated in South America, and the Weft India iflands. It rifes to fifty or fixty feet high, the body or trunk generally leaning on one fide; but is regularly fhaped, being equally thick at both ends, and fmalleft in the middle. The bark is fmooth, and of a pale brown colour. At the top come out from twenty to thirty branches, or rather leaves, fome of them fifteen feet long; thefe

are

are winged, ftraight, and tapering. The
lobes are green, fword-fhaped, and about
three feet long towards the bafe of the midrib, but diminifh towards the extremity.
The branches or leaves are bound at their
bafe by ftringy threads, about the fize of
fmall packthread, which are interwoven like
a web. The flowers are of a pale yellow
colour, are produced in long bunches at the
infertions of the leaves, and are male and
female iffuing from the fame fheath. The
male is compofed of a fmall, three-leaved
calyx, containing three oval, fharp-pointed
petals, and fix ftamina, tipped with arrow-
fhaped fummits. The female alfo has a
three-leaved calyx, and three petals, furrounding one ftyle, crowned by a three-
lobed ftigma. The germen is oval, and
fwells to a large berry, inclofing an oval nut,
with a hard fhell, having three holes at the
top, and is covered with a kind of tow,
which the Indians twift off, and make into
cordage. With this tow they likewife make
an excellent caulking for their veffels.

Within the Nut is found a kernel, as
pleafant as an Almond, and alfo a large
quantity of liquor refembling milk, which
the Indians greedily drink before the fruit is
ripe, it being then pleafant, but when the
Nut is matured, the liquor becomes four.
Some full-grown Nuts will contain a pint or
more of this milk, the frequent drinking of

which

which feems to have no bad effects upon the
Indians, yet Europeans fhould be cautious
of making too free with it at firft, for when
Lionel Wafer was at a fmall ifland in the
South Sea, where this tree grew in plenty,
fome of his men were fo delighted with it,
that at parting they were refolved to drink
their fill, which they did; but their appe-
tites had like to have coft them their lives,
for though they were not drunk, yet they
were fo chilled and benumbed, that they
could not ftand, and were obliged to be
carried aboard by thofe who had more pru-
dence than themfelves, and it was many
days before they recovered.

The fhells of thefe Nuts being hard, and
capable of receiving a polifh, they are often
cut tranfverfely, when being mounted on
ftands, and having their edges filvered or
gilt, or otherwife ornamented, thus ferve
the purpofe of drinking cups. The leaves
of the tree are ufed for thatching, for brooms,
bafkets, and other utenfils; and of the re-
ticular web, growing at their bafe, the Indian women make cauls and aprons.

6 FAGUS caftanea. *Common Chefnut.*
Lin. Sp. pl. 1416.

Caftanea fylveftris. *Bauh Pin.* 419.

The *Common Chefnut* is a native of the
fouthern parts of Europe, but is much cultivated in England, where it produces as
good

good fruit as it does in Spain and Italy, though they are not altogether fo large. It is now fo common in gentlemens plantations, that it is generally known. It will be needlefs to fpeak about the nature of the Nuts, but it may be obferved, that the tree affords excellent timber, the wood being equal to the beft Oak for many purpofes.

7 FAGUS pumila. *American Chefnut.* *Lin. Sp. pl.* 1416.

Fagus foliis ovato-lanceolatis ferratis. *Roy. lugdb.* 79.

This is a native of America. It differs from the former in the tree being much fmaller; in the leaves being woolly underneath, and in the catkins of flowers being flenderer and knotted. The Nuts are a little bigger than Hazel-nuts, but far exceed the Common Chefnut in fweetnefs. The woods of South Carolina abound with thefe trees.

8 JUGLANS regia. *Common Walnut*. *Lin. Sp. pl.* 1415.

Nux juglans five regia vulgaris. *Bauh. Pin.* 417.

The *Common Walnut* is known to all by being fo univerfally cultivated, but its native place of growth has not yet been afcertained. There are many varieties of it, which are only feminal variations. The

A a 4 meats

meats are fuppofed to be much of the nature of Almonds, yet they are certainly lefs emollient, as they are apt to excite coughing. The Chinefe candy thefe Nuts into a Sweetmeat, and the raw kernels they put into their tea, as has been mentioned of the Hazel-nuts.

9 JUGLANS nigra. *Black Virginian Walnut. Lin. Sp. pl.* 1415.

This grows naturally in Virginia and Maryland, where it arrives to a large fize, having its branches furnifhed with leaves, compofed of five or fix pair of fpear-fhaped lobes; thefe are ferrated, fharp-pointed, and the lower pair the leaft. When rubbed they emit a ftrong aromatic fmell, as do alfo the Nuts, which are rough, rounder than the Common Walnut, their fhells very hard and thick, the kernels fmall, but fweeter than our nuts.

10 JATROPHA curcas. *Indian Phyfic Nut. Lin. Sp. pl.* 1429.

Jatropha affurgens, ficûs folio, flore herbaceo. *Browne's Jam.* 348.

This grows naturally in the Weft India iflands, where it rifes with a ftrong ftem to about fourteen feet, divided into feveral branches, furnifhed with angular heartfhaped leaves, fomewhat refembling thofe of the Fig. The flowers are male and fe-

male

male diftinct on the fame plant, of an herbaceous colour, and are produced in umbels
at the ends of the branches. The females
are fucceeded by oblong-oval capfules, with
three cells, each containing one oblong
black feed.

11 JATROPHA multifida. *French Phyfic
Nut. Lin. Sp. pl.* 1429.

Jatropha affurgens, foliis digitatis: la-
ciniis anguftis pinatifidis. *Browne's Jam.*
348.

This is a native of South America, but
is cultivated in the Weft Indies. It is a
lower fhrub than the former, and the leaves
are divided into nine or ten narrow lobes,
which are joined at·their bafe, and have
many jagged teeth on their edges, ftanding
oppofite. The upper furface of the leaves
are of a fhining green, but the under fide
greyifh. The flowers are male and female
diftinct on the fame plant, and of a bright
fcarlet colour; they come out in umbels in
manner of the former, and make a beautiful
appearance, whereby the fhrub is as much
cultivated for ornament, as for ufe.

The kernels of the Nuts of both thefe
fpecies are violently emetic and cathartic, as
many European failors have experienced;
for only three or four of them, eaten by
people ignorant of the Nuts, and the effects
of the kernels, have purged them both ways

for

for many hours after. The natives affirm
that this purgative quality confifts entirely
in a film that runs through the centre of
the kernel ; and Dr. Bancroft fays he really
believes this to be the cafe, he having frequently eaten the meats when divefted of
this membrane, without feeling any of the
above effects. The kernels have a grateful
flavour.

12 PINUS pinea. *Stone Pine. Lin. Sp.*
pl. 1419.

Pinus officulis duris, foliis longis. *Bauh.*
*Hift.,*I. *p.* 248.

This is a large tree, and grows naturally
in France, Spain, and Italy. The leaves
grow two in a fheath, are a little ciliated,
inclining to a fea-green colour, and are rather thinner and fhorter than thofe of the
Pineafter. The cones are roundifh, very
thick, about five inches long, and the fcales
end in an obtufe point. The feeds are near
three quarters of an inch long, thick, inclining to an oval form, round backed, and
of a light brown colour.

The kernels of thefe Nuts or feeds have
a pleafant, agreeable tafte, and in Italy are
frequently ferved up in deferts. An oil is
drawn from them, which is equal in goodnefs to that obtained from Hazel - nuts.
Between the wood and inner bark of this
tree, lies a foft white fubftance, which in
the

the Spring the Swedes prepare a much-
efteemed dilh from; and the bark is often
ground and mixed with Oat-meal for bread.

13 PISTACIA vera. *Piftachia Nut.* *Lin.*
Sp. pl. 1454.

Piftacia peregrina, fructu racemofo five
Terebinthina indica. *Bauh. Pin.* 401.

The *Piftachia* grows in feveral parts of
Afia. It rifes to between twenty and thirty
feet; the young branches are covered with
a light-brown bark. The leaves are pinnated, and compofed of about three pair of
oval lobes, with an odd one at the end.
The lobes emit an odour on being rubbed,
and their edges are turned backwards. It
hath male and female flowers in diftinct
plants. The males are produced in loofe
fparfed catkins. They have no petals, but
each confifts of a fmall five-pointed calyx,
containing five fmall ftamina, terminated by
four-cornered fummits. The female flowers
come out in clufters from the fides of the
branches; thefe have no petals, but each
has a large oval germen, fupporting three
reflexed ftyles, and are fucceeded by oval
Nuts.

The kernels of thefe Nuts have a fweet,
unctuous tafte, refembling that of fweet
Almonds. They are of a healing balfamic
nature, and are deemed ferviceable in diftempers of the breaft.

14 PISTACIA

14 PISTACIA narbonenfis. *Trifoliate-leaved Turpentine-tree. Lin. Sp. pl.* 1454.

Terebinthus indica major, fructu rotundo. *Bauh. Hift.* I. *p.* 277.

This grows naturally in Perfia, and fome parts of Armenia. It is a middling-fized tree, fending out many fide branches, fur-nifhed with light-green winged leaves, com-pofed of three or five roundifh lobes, ftand-ing on long footftalks. It is male and fe-male in diftinct plants, as the former. The Nuts are fmall, but their kernels are eaten in manner of the true fort.

15 THEOBROMA cacao. *Chocolate Nut. Lin. Sp. pl.* 1100.

Amygdalis fimilis guatimalenfis, *Bauh. Pin.* 442.

The *Chocolate Nut-tree* grows naturally upon moft parts of the ifthmus of Darien, and feveral of the Spanifh fettlements in the Weft Indies. It rifes to a confiderable height in its natural ftate, but when cul-tivated for a crop, it is topped to keep it low. The leaves are very large, oval, and entire. The flower is compofed of five flefh-coloured petals, which are irregularly in-dented, and furround five erect, awl-fhaped ftamina, and one like fhaped ftyle, crowned with a fimple ftigma. The germen is nearly oval, and becomes a yellow oblong pod, about the fize of a Melon, pointed at both ends,

ends, and having five cells, filled with oval, compreffed, flefhy feeds.

Thefe feeds or Nuts are about the fize of Olives, are of an oily nutritive nature, and conftitute a principal part of what is fold in the fhops by the name of *Chocolate.*

In order to cure the Nuts for fale, the negroes cut the pods lengthways, and take them out, at the fame time carefully divefting them of the pulp which fticks about them. This done, they are carried to a houfe, and laid in large wooden veffels raifed above the ground, when they are covered with mats, upon which they place boards with weights upon them, to prefs the Nuts clofe. In thefe veffels they are kept to ferment for four or five days, but they muft be well ftirred every morning, left the exceffive heat fhould fpoil them, and in the end they change from a white to a brown colour. Afterwards they are taken out of the veffels, fpread upon cloths, and expofed in the fun to dry, and when fufficiently weathered, they are packed up for market.

16 TRAPA natans. *Jefuit's Nut. Lin. Sp. pl.* 175.

Tribulus aquaticus. *Bauh. Pin.* 194.

This grows plentifully in the lakes and ftagnant waters in Italy and Germany. It hath almoft femicircular leaves, which float

6

on the furface of the water; among which rife up fappy, round ftalks, fupporting the flowers. Each flower hath a monophyllous calyx, cut into four acute parts, and fur-rounds four oval, whitifh petals, larger than the calyx. In the centre are four ftamina, and one ftyle, crowned with a roundifh fnipped ftigma. The germen is oval, and becomes a naked oblong-oval Nut, having one cell, and armed with four fharp, thick-ifh fpines, ftanding oppofite one another in the middle.

Thefe Nuts are collected by the common people, and their kernels having a pleafant flavour, are not only eaten crude, but are often made into bread.

C H A P.

C H A P. XI.

ESCULENT FUNGUSES.

1 **A**GARICUS campeſtris. *Common Muſhroom.*
2 Agaricus pratenſis. *The Champignion.*
3 Agaricus chantarellus. *Chantarelle Agaric.*
4 Agaricus deliciofus. *Orange Agaric.*
5 Agaricus cinnamomeus. *Brown Muſh-room.*
6 Agaricus violaceus. *Violet Muſhroom.*
7 Lycoperdon tuber. *The Truffle.*
8 Phallus efculentus. *The Morel.*

As the Agarics are numerous, and generally ſuppoſed to be poiſonous, I ſhall deſcribe the above few wholeſome ones as minutely as poſſible, in order to prevent any accident from miſtaking the ſpecies.

1 AGARICUS campeſtris. *Common Muſh-room. Lin. Sp. pl.* 1641.
Fungus campeſtris albus ſupernè, infernè rubens. *Bauh. Hiſt.* III. *p.* 824.
The top or cap of this is firſt of a dirty cream colour, convex, and if but juſt ex-
panding,

panding, the under part, or what is called the gills, is of a bright flesh red; this colour lasts but a little time before it turns darker; and when the plant is old, or has been some time expanded, the gills become of a dark brown, the cap almost flat, of a dirty colour, and often a little scaly. It differs much in size, in different plants, it being from an inch to seven inches broad. The general use of it is well known. It is found in woods, old pastures, and by road sides, and is in the greatest perfection in September.

There is a variety of this with a yellowish white cap and white gills; this is very firm, but seldom expands so freely as the true sort, and when broiled will exude a yellowish juice. It is probable this sort is not pernicious, though it is always rejected by such as can distinguish it.

2 AGARICUS pratensis. *Champignion.* *Hudson's Flo. Angl.* 616.

The *Champignion* is very common upon heaths and dry pastures. A number of them generally come up in a place, ranged in curved lines or circles. The cap is small, almost flat, from one to two or three inches diameter, of a pale buff colour, often crimpled at the edges, and when dry, tough like leather, or a thin piece of fine cork. The gills are of the colour of the cap, are thinly placed, with a short one, and sometimes two,

coming

coming from the edge of the cap between each. The stalk or pillar is also of the colour of the cap; it is long, slender, and all the way of a thickness.

This plant has but little smell, is rather dry, and yet when broiled or stewed, it communicates a good flavour. In perfection with the former.

3 Agaricus chantarellus. *Chantarelle Agaric. Lin. Sp. pl.* 1639.

Fungus minimus flavescens infundibuli-formis. *Bauh. Pin.* 373.

This is rather a smaller Fungus than the former. The cap is yellow, of different hues in different plants, some being of a pale yellow, and others of an orange colour. It is generally sunk in the middle, somewhat resembling a tunnel, and its edges are often twisted and contorted so as to form sinuses or angles. The gills are of a deeper colour than the outside, are very fine, even, numerous, and beautifully branched. The ramifications begin at the stalk, and are variously extended towards the edge of the cap. The pillar is of the same colour as the cap, is seldom inserted in the centre, but rather sideways; it is short, thickish at the root, and the gills mostly run down the top, which make it appear smallest in the middle.

This plant broiled with salt and pepper
has

has much the flavour of a roasted cockle; and is esteemed a delicacy by the French, as is the former. It is found in woods and high pastures, and is in perfection about the end of September.

4 AGARICUS deliciosus. *Orange Agaric.*
Lin. Sp. pl. 1641.
Amanita fulvus, lacte croceo. *Hall. Hist.* 2419.

The general size of the cap of the *Deliciosus* is from two to four inches broad. Its form is circular, with the edges bent inwards; convex on the upper surface, except in the centre, where it is a little depressed, so as nearly to resemble the apex of a smooth Apple. The colour is a sordid yellow, streaked with ash and yellowish brown, from the centre to the edge, and when it is broken, it emits a gold-colour juice. The gills are of a deep yellow, and a few of them come out by pairs at the stalk, but divide immediately, and run straight to the edge of the cap. The stalk or pillar is thinnest near the middle, thickest at the root, and when cut transversely, it is quite white in the centre, with a fine yellow ring that goes to the edge.

This Fungus well seasoned and then broiled, has the exact flavour of a roasted Muscle. Its prime time is September, and it is to be found in high dry woods.

5 AGARICUS

5 Agaricus cinnamomeus. *Brown Mushroom. Lin. Sp. pl.* 1642.

The *Brown Mushroom* has a cap the colour of fresh-tanned hides. At first it is hemispherical, firm, even, and fleshy, with mostly a small rising in the centre; but when old it is quite flat, and then somewhat resembles the *lactifluus*, except that it is not milky. The gills are of a yellowish brown, not very distant from each other, bent like a knee at the pillar, and have a short one or two run from the edge of the cap between each. The pillar is near the length of a finger, firm, rather thick, brown, at the base, of a sordid yellow upward, and when cut transversely, of a fine white grain. The cap in different plants is from two to five inches broad.

The whole plant has a pleasant smell, and when broiled gives a good flavour. It is found in woods, in September and October.

6 Agaricus violaceus. *Violet Mushroom. Lin. Sp. pl.* 1641.

Fungus esculentus bulbosus dilutè purpureus. *Mich. Gen.* 149. *t.* 49. *f.* 1.

The cap of the *Violet Mushroom*, when first expanded, is smooth, hemispherical, the main surface of a livid colour, but towards the margin it is of a better blue. When full grown or old it becomes corrugated, and of a rusty brown. The gills of

a young

a young plant are of a beautiful violet colour, and regularly placed. The pillar is of the colour of the gills, fhort, of a conical form, but fwelled at the bafe into a fort of bulb. Its upper part is furrounded with an iron-coloured wool, which, in a plant juft expanding, ftretches crofs to the edge of the cap like a web.

This requires much broiling, but when fufficiently done and feafoned, it is as delicious as an Oyfter. It is found in woods in October, and I have met with plants from two to fix inches broad. Hudfon's *bulbofus* is only a fmall variation of this plant.

There are fome other fpecies of Agarics that are frequently eaten by the country people; and it is probable the greateft part of thofe with firm flefhy caps might be eaten with fafety, provided they were chofen from dry grounds. It is well known that foil and fituation have a great influence upon the properties of plants; and thefe being of a fingular nature, and abfolutely between that of an animal and vegetable, may be more powerfully affected than a compleat fpecies of either, by reafon they have neither leaves nor branches to carry off the noxious damps and vapours of a ftagnant foil, as a perfect vegetable has; nor have they any grofs excremental difcharges, like thofe of a living animal. The gills no doubt do exhale fome of their fuperfluous moifture, but their

fituation

fituation is fuch, that any thick fteam from
the earth may lodge in them, and by clog-
ging their excretory ducts, render the plants
morbid. Thus they foon run into a ftate of
putrefaction, and become a prey to worms,
flies, and other infects. The common
Mufhroom, which is in general efteem,
(though we have feveral others better) is
not fafely eaten, when produced upon a
moift foil. An acquaintance of mine, who
is exceeding fond of broiled flaps, as he calls
them, was taken very ill upon eating fome
he gathered off a wet cloggy land. He became very fick, with his ftomach much
diftended, which induced him to think he
was abfolutely poifoned; but luckily for
him, he had fome fat mutton broth in the
houfe, of which he drank plentifully, and
his ftomach difgorging, he recovered. This
accident, however, did not difcourage him
from making free with his beloved difh in
future, but he has been careful ever fince to
gather his Mufhrooms (and no one knows
Mufhrooms better) on dry foils; being
himfelf convinced, that the pernicious
quality of his flaps, was entirely owing to
the place they grew upon.

From this it is evident, that thofe who
gather Mufhrooms for fale, fhould have particular regard to the lands they collect them
from, efpecially if they know they are to be
broiled; but if they be intended for *Catchup,*

perhaps

perhaps they may be lefs cautious, as the
falt and fpices, with which the juice is
boiled, may correct any evil difpofition in
the plants. But even in this cafe, I can
from my own experience aver, that *Catchup*
made of Mufhrooms taken from a dry foil,
has a more aromatic and pleafant flavour,
than that which is made of thofe taken
from a moift one, and it will always keep a
great deal better.

7 LYCOPERDON tuber. *The Truffle.*
Lin. Sp. pl. 1653.

Tuber brumale, pulpa obfcura odorata.
Mich. Gen. 221. *t.* 164.

The *Truffle* is a folid Fungus, of a globular figure, and grows under the furface of
the ground, fo as to be totally hidden. It
has a rough blackifh coat, and is deftitute
of fibres. The manner of its propagation
is entirely unknown. Cooks are well acquainted with its ufe and qualities. It is
found in woods and paftures in fome parts
of Kent, but is not very common in England. In France and Spain *Truffles* are very
frequent, and grow to a much larger fize
than they do here. In thefe places the
peafants find it worth their while to fearch
for them, and they train up dogs and fwine
for this purpofe, who after they have been
inured to the fmell, by their mafters frequently placing fome in their way, will
                                    readily

readily fcrape them up as they ramble the fields and woods.

8 Phallus efculentus.   *The Morel. Lin. Sp. pl.* 1648.

Boletus capite tereti reticulato.   *Hall. Hift.* 2247.

The *Morel* is a Fungus of a very fingular conftruction, having an oval, or rather co-nical head, full of irregular pits or cells, which in the larger plants are big enough to receive the tip of a finger.   The centre of the bafe is faftened to a thick ftalk, about the length of the head, and irregularly fluted near the root.   The whole plant at firft is nearly of a buff colour, but when old it becomes brown.   It grows on moift banks, and wet paftures, and fprings up in May. It is ufed in the fame manner as the Truffle for gravies, but gives an inferior flavour.

B b 4   APPENDIX.

APPENDIX.

THE following plant could not with propriety come under any of the general divisions of the foregoing work.

HIBISCUS efculentus. *Fig-leaved Okra.*
Lin. Sp. pl. 980.

Alcea maxima, malvæ rofeæ folio, fructu decagono recto craffiore breviore efculento. *Browne's Jam.* 284. *n.* 3.

This is an annual, and a native of both the Indies. It fends up a fpungy ftalk rather more than a yard high, which branches towards the top, and is furnifhed with hand-fhaped leaves, having five lobes. The flowers are produced at the divifions of the ftalk; each has a double calyx, and the under one is torn on one fide. The petals are heart-fhaped, are five in number, of a fulphur colour, are joined at their bafe, and have dark purple bottoms. The ftamina are many, and are united into a column below, but expand near the top. The germen is roundifh, and turns to a thick capfule, three or four inches long, moftly ftanding erect, and having five cells, containing kidney-fhaped feeds.

The inhabitants of the Indies boil thefe

6   pods

pods in their foups. They contain a vifcous acid juice, which communicates a thicknefs, and alfo a pleafant flavour.

The generic characters of the following two species have not yet been perfectly settled.

GINKGO. *Maiden-hair Tree.*
Arbor nucifera, folio adiantino. *Kæmpf. Amœn. Exot.* 811.

This is a native of Japan, where it is known by the names *Ginan* and *Itſio.* The body is covered with an afh-coloured bark, and a full-grown tree is as large as a Walnut. The wood is brittle, having a foft fpungy pith running through it. The leaves are large, and expand in the form of a Maiden-hair leaf. They are narrow at the bafe, unequally divided upward, have no nerves or fibres, both furfaces being alike. The upper fide of the footftalk is flat, and runs into the fubftance of the leaf. The flowers are produced in long catkins, at the bofoms of the leaves of the young twigs, and are fucceeded by plums, nearly of the fize and colour of the Damafk Plum, each containing a whitifh, brittle ftone, refembling that of the Apricot, but larger, enclofing a white kernel, having much the flavour of an Almond.

In China and Japan thefe kernels always make part of the defert at all public feafts
and

and entertainments. They are faid to promote digeftion, and to cleanfe the ftomach and bowels.

BREAD FRUIT-TREE.

This grows in all the Ladrone Iflands in the South Sea, as is mentioned by Capt. Dampier and Lord Anfon, and alfo at Otaheite, by Capt. Cooke, and is thus defcribed:

The *Bread Fruit* grows on a tree about the fize of a middling Oak. Its leaves are a foot and half long, of an oblong figure, deeply finuated like thofe of the Fig-tree, which they refemble in confiftence and colour, and in exuding a milky juice upon being broken. The fruit is about the fize of a child's head, and the furface is reticulated, not much unlike a Truffle. It is covered with a thin fkin, and has a core about the fize of a fmall knife. The edible part is between the fkin and the core; it is as white as fnow, and fomewhat of the confiftence of new bread. It muft be roafted before it is eaten, being firft divided into three or four parts. Its tafte is infipid, with a flight fweetnefs, nearly like that of wheaten bread, mixed with Jerufalem Artichoke.

This Fruit is the conftant food of the inhabitants all the year, it being in feafon eight months; and in order to fupply the remaining

remaining four, they have a method of fweating the unripe fruit, by laying them in heaps in a hole made in the floor of the houfe (which hole they neatly line with grafs) and covering them with leaves, and a layer of ftones, by which they ferment and become four, and will then keep for feveral months. This mafs is called *Mahie*, and as it is wanted, it is taken out of the hole, made into balls, wrapped in leaves, and baked.

I N D E X

INDEX

OF

LATIN NAMES.

I N D E X.

Fragaria

INDEX.

Phalaris

INDEX.

INDEX

INDEX

OF

ENGLISH NAMES.

INDEX.

INDEX.

INDEX.

INDEX.

3

I N D E X.

N. B. *The Author not having an opportunity of seeing the sheets, till after they were worked off, finds it necessary to correct the following*

E R R A T A.

Page 17, l. 12, for fparedly, r. fparfedly.
—— 18, l. 28, take away the comma between rats and granaries.
—— 35, l. 4, for cut, r. eat.
—— 46, l. 8, for quantities, r. qualities.
—— 99, l. 27, and
—— 115, l. 2, for fpikes, r. racemi.
—— 101, l. 21, for hirfutia, r. hirfutie.
—— 182, l. 22, and where elfe the expreffion occurs, for thefe fruits, r. the fruit.
—— 189, l. 9, for apples, r. berries.
—— 193, l. 26, for plums, r. berries.
—— 308, l. 17, for rind, r. rinded.
—— 377, l. 1, for pods, r. capfules.
—— ibid. l. 12, for walnut, r. walnut-tree.

www.ingramcontent.com/pod-product-compliance
Lightning Source LLC
Chambersburg PA
CBHW021350210326
41599CB00011B/821